Rainer Glüge

Homogenization Methods

Also of Interest

Homogenisierungsmethoden.
Effektive Eigenschaften von Kompositen
Rainer Glüge, 2021
ISBN 978-3-11-071948-2, e-ISBN (PDF) 978-3-11-071949-9,
e-ISBN (EPUB) 978-3-11-071952-9

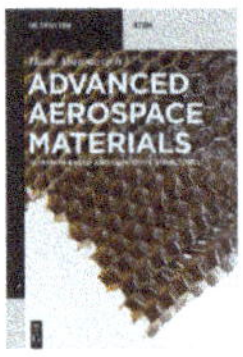

Advanced Aerospace Materials.
Aluminum-Based and Composite Structures
Haim Abramovich, 2019
ISBN 978-3-11-053756-7, e-ISBN (PDF) 978-3-11-053757-4,
e-ISBN (EPUB) 978-3-11-053763-5

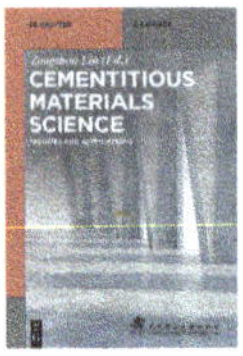

Cementitious Materials Science.
Theories and Applications
Lin Zongshou (Ed.), 2019
ISBN 978-3-11-057209-4, e-ISBN (PDF) 978-3-11-057210-0,
e-ISBN (EPUB) 978-3-11-057216-2

Tensor Analysis
im Maschinen- und Fahrzeugbau
Heinz Schade, Klaus Neemann, 2018
ISBN 978-3-11-040425-8, e-ISBN (PDF) 978-3-11-040426-5,
e-ISBN (EPUB) 978-3-11-040549-1

Geometry of Incompatible Deformations.
Differential Geometry in Continuum Mechanics
Series: De Gruyter Studies in Mathematical Physics, 50
Sergey Lychev, Konstantin Koifman, 2019
ISBN 978-3-11-056201-9, e-ISBN (PDF) 978-3-11-056321-4,
e-ISBN (EPUB) 978-3-11-056227-9

Rainer Glüge

Homogenization Methods

Effective Properties of Composites

DE GRUYTER
OLDENBOURG

Author
Priv.-Doz. Dr.-Ing. habil. Rainer Glüge
gluege@boerde.de

ISBN 978-3-11-079351-2
e-ISBN (PDF) 978-3-11-079352-9
e-ISBN (EPUB) 978-3-11-079365-9

Library of Congress Control Number: 2022948647

Bibliographic information published by the Deutsche Nationalbibliothek
The Deutsche Nationalbibliothek lists this publication in the Deutsche Nationalbibliografie; detailed bibliographic data are available on the Internet at http://dnb.dnb.de.

Cover image: asbe / iStock / Getty Images Plus
Typesetting: VTeX UAB, Lithuania
Printing and binding: CPI books GmbH, Leck

www.degruyter.com

Preface

For my Ph. D. thesis, I was concerned with modeling deformation twinning in single crystals. These are shear bands of varying thicknesses and fixed orientation with respect to the crystal lattice. To obtain effective, macroscopic stress-strain relations, I used a periodically repeated cubic unit cell that covered some 30 grains. I soon realized that the ability of the unit cell to accommodate deformation twins depends on the alignment of the periodicity frame with the crystal orientations, which affects the effective material response. This erroneous orientation dependence raised my curiosity about homogenization methods. Without being aware of the vastness and complexity of the subject, I started exploring it.

After teaching homogenization methods for several years, the lecture notes matured into the German edition of this book, of which this is an updated translation. I have included many minor corrections and a few additional sections.

The book covers a wide range of homogenization methods for linear problems. I did my best to maintain a complete overview and point out the connections between different approaches.

To check for myself whether I have understood something, I found it helpful to implement it as a computer program. Gaps and errors in reasoning emerge mercilessly when putting them into an algorithm. As a result, the book contains many listings of Mathematica notebooks,[1] which may be useful for practitioners.

This book cannot be digested without a firm basis in continuum mechanics and tensor calculus.

1 Available at https://gitlab.com/gluegerainer/listings-homogenization-methods

https://doi.org/10.1515/9783110793529-201

Contents

1 Introduction

Homogenization is the calculation of the effective properties of inhomogeneous materials. It requires as input the material arrangement and the local material properties on the microscale, while the effective material properties are provided on a larger scale, involving a scale transition.

For example, metal parts are composed of numerous single crystal grains with anisotropic constitutive laws. If the orientation distribution is homogeneous, one can assume effective isotropic material on the macroscale, and these can be calculated with the methods of homogenization.

One may be tempted to assume that the effective material properties are obtained by simply volume averaging the local material properties, as it is, for instance, possible for the mass density. The effective mass density can be obtained by multiplying the partial mass densities with the volume fractions v_i,

$$\bar{\rho} = \sum_{i=1,n} v_i \rho_i, \quad \sum_{i=1,n} v_i = 1. \tag{1.1}$$

This is only possible when the spatial arrangement of the phases does not matter. For elasticity, thermal or electrical conductivity, and other properties, this can lead to bad estimates of the effective properties. For example, when mixing isotropic phases, the arrangement may be anisotropic, leading to effectively anisotropic properties that cannot be obtained by averaging isotropic quantities. But even if the microstructure is effectively isotropic, the phases may be arranged as a matrix-inclusion structure or as interpenetrating phases, resulting in largely different isotropic properties.

Moreover, the interaction of the phases on the microscale can give rise to effective properties that are not present in any of the phases. As already mentioned, the local arrangement of isotropic phases can result in an effective anisotropy. Another example is precipitation hardening: If an ideally plastic material without hardening contains stiff precipitates, the effective plastic behavior shows pronounced hardening. The appearance of phenomena on the macroscale that can be approximated with fewer equations and degrees of freedom than the entire microstructure with the interacting phases is termed emergence.

The term may be associated therefore with a reduction of complexity (replacing a microstructure with many degrees of freedom with a macromodel with fewer degrees of freedom) as well as with an increase in complexity (appearance of additional phenomena), depending on whether you find number crunching complex or constitutive equations. Emergence is therefore the opposite of fundamental, as the emergent phenomena can be explained by more fundamental, underlying modeling. Emergent phenomena cannot be described by simple averaging of the fundamental properties on the microscale.

https://doi.org/10.1515/9783110793529-001

The task of homogenization is to provide simple and precise models with few degrees of freedom that predict the effective material properties on the macroscale, basically reducing a large amount of microscale information.

This book gives an overview of the broad field of homogenization with application to linear continuum mechanics. Figure 1.1 shows one way to structure the methods presented here. The intuitively easy-to-understand representative volume element (RVE) problem may be a good starting point. In Chapter 5.1, the effective elastic properties of a honeycomb structure are derived with the aid of beam theory, which is even accessible without continuum mechanics. But an important class of methods is only obtained after reformulating the RVE problem as the difference problem, which has advantageous properties for analytical treatment.

Literature

I recommend the following books for getting a more complete view of the topic:

- G. W. Milton (2002). *The Theory of Composites*. Cambridge University Press.
- A. Morawiec (2004). *Orientations and Rotations – Computations in Crystallographic Textures*. Springer.
- D. Gross and T. Seelig (2011). *Fracture Mechanics – With an Introduction to Micromechanic*. Springer.
- R. M. Brannon (2018). *Rotation, Reflection, and Frame Changes*. IOP Publishing.

The books of Milton and Morawiec have a more formal, mathematical character. Milton's book lends itself as a work of reference. The books of Brannon, Gross and Seelig address engineers. They also contain an introduction to continuum mechanics, which is presumed here. Further, the book

- S. Nomura (2016). *Micromechanics with Mathematica*. Wiley

is useful as a preparation for this book, since it contains an introduction to linear continuum mechanics as well as to the computer-aided math system, Mathematica.

Although I did my best to weed out errors, the book contains probably some mistakes. I am grateful for corrective notes to `gluege@boerde.de`.

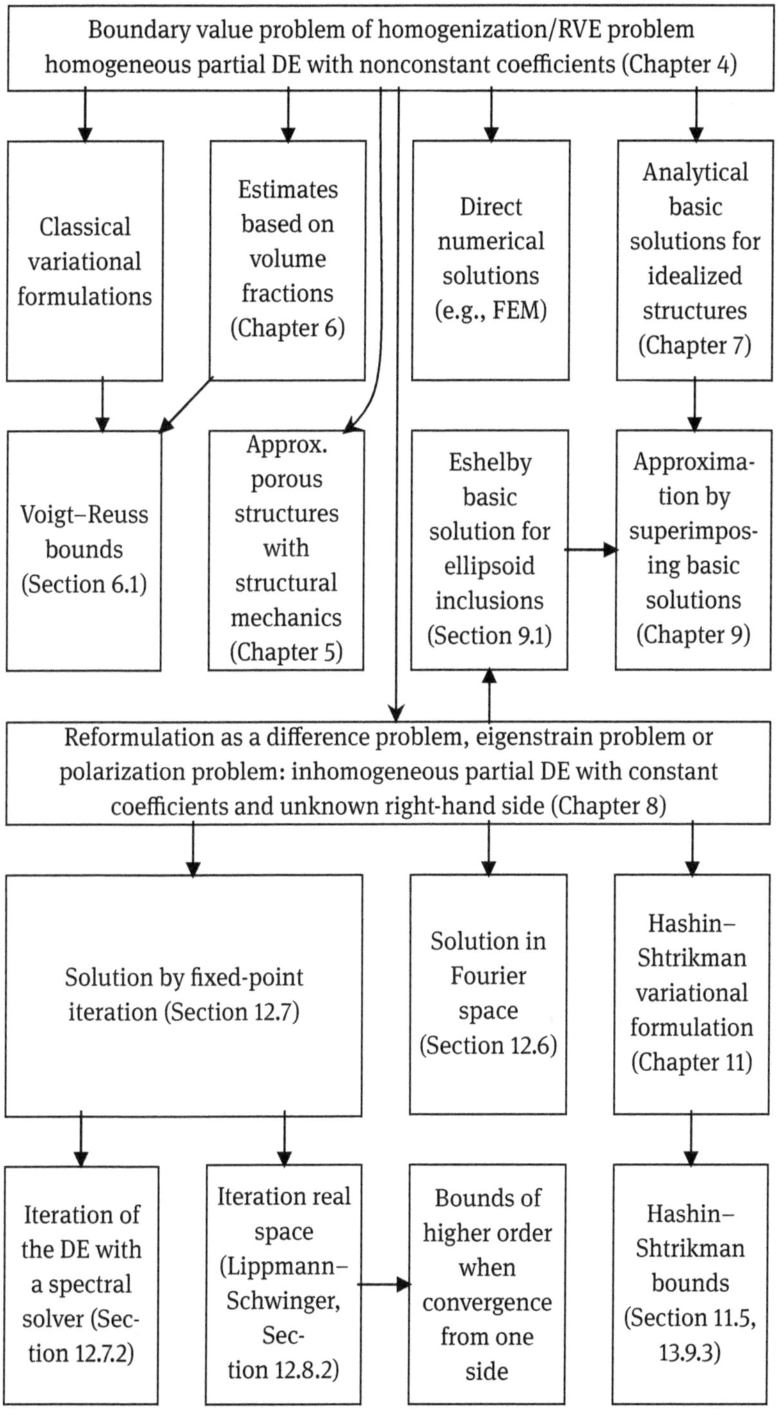
Boundary value problem of homogenization/RVE problem
homogeneous partial DE with nonconstant coefficients (Chapter 4)
Classical variational formulations
Estimates based on volume fractions (Chapter 6)
Direct numerical solutions (e.g., FEM)
Analytical basic solutions for idealized structures (Chapter 7)
Voigt–Reuss bounds (Section 6.1)
Approx. porous structures with structural mechanics (Chapter 5)
Eshelby basic solution for ellipsoid inclusions (Section 9.1)
Approximation by superimposing basic solutions (Chapter 9)
Reformulation as a difference problem, eigenstrain problem or polarization problem: inhomogeneous partial DE with constant coefficients and unknown right-hand side (Chapter 8)
Solution by fixed-point iteration (Section 12.7)
Solution in Fourier space (Section 12.6)
Hashin–Shtrikman variational formulation (Chapter 11)
Iteration of the DE with a spectral solver (Section 12.7.2)
Iteration real space (Lippmann–Schwinger, Section 12.8.2)
Bounds of higher order when convergence from one side
Hashin–Shtrikman bounds (Section 11.5, 13.9.3)

Figure 1.1: Structure of homogenization methods.

2 Basics

2.1 Notation

The symbolic notation here is a compromise between different styles. Vectors are denoted as small bold letters, (e. g., $\mathbf{x}$) and tensors of order two as capital bold letters (e. g., $\mathbf{E}$). Only the stress and strain tensors $\boldsymbol{\sigma}$ and $\boldsymbol{\varepsilon}$ are an exception. Higher-order tensors are written in blackboard bold capital letters (e. g., $\mathbb{C}$). The order is denoted above if necessary.

In the component-basis representation, orthonormal bases are used with a few exceptions. The index notation implies summation over indices that occur twice in a product, e. g., $\mathbf{v} = \sum_{i=1\ldots3} v_i\mathbf{e}_i = v_i\mathbf{e}_i$. Locations in $\mathbb{R}_3$ are denoted with components and basis, $\mathbf{x} = x_i\mathbf{e}_i$. In curved spaces like SO(3) (Section 13), one can only give coordinates.

The term "linear" is used in the sense of algebra. A mapping f is linear if

$$f(\alpha\mathbf{a} + \mathbf{b}) = \alpha f(\mathbf{a}) + f(\mathbf{b}) \tag{2.1}$$

holds. In the sense of analysis, a mapping of the form $y(x) = mx + n$ is linear. The latter function is only linear in the sense of algebra if $n = 0$.

The dyadic product and single scalar contractions are denoted as follows: $(\mathbf{a} \otimes \mathbf{b} \otimes \mathbf{c}) : (\mathbf{d} \otimes \mathbf{e}) = (\mathbf{b} \cdot \mathbf{d})(\mathbf{c} \cdot \mathbf{e})\,\mathbf{a}$. The scalar associations are such that the n-fold scalar product of two tensors of order n is positive definite, e. g., for second-order tensors:

$$\mathbf{A} : \mathbf{A} = A_{ij}A_{kl}\mathbf{e}_i \otimes \mathbf{e}_j : \mathbf{e}_k \otimes \mathbf{e}_l = A_{ij}A_{kl}\delta_{ik}\delta_{jl} = A_{ij}A_{ij} = \sum_{i,j=1\ldots3} A_{ij}^2 = \|\mathbf{A}\|^2 > 0 \tag{2.2}$$

$$\text{with} \quad \mathbf{e}_i \cdot \mathbf{e}_k = \delta_{ik} = \begin{cases} i = k : 1 \\ i \neq k : 0. \end{cases} \tag{2.3}$$

The arrangement of the scalar dots above or beside each other does not matter. The scalar dot can be omitted if the scalar association is clear from the context, e. g., in $\mathbf{a} = \mathbf{A}\mathbf{b}$. The abbreviation δ_{ij} is the Kronecker symbol. It can be used to denote the second-order identity tensor,

$$\mathbf{I} = \delta_{ij}\mathbf{e}_i \otimes \mathbf{e}_j = \mathbf{e}_i \otimes \mathbf{e}_i = \mathbf{e}_1 \otimes \mathbf{e}_1 + \mathbf{e}_2 \otimes \mathbf{e}_2 + \mathbf{e}_3 \otimes \mathbf{e}_3, \tag{2.4}$$

or the orthonormality of the basis $\mathbf{e}_i$ as in equation (2.3). Since a product involving δ_{ij} is only nonzero when $i = j$, one can simplify on such occasions as in equation (2.2).

Transpositions are interchanges of the inputs of multilinear forms or tensors. For second-order tensors, the swapping of the two inputs is indicated by a superscript T. In the case of orthonormal bases, this can be reduced to simple index interchanges, e. g.,

https://doi.org/10.1515/9783110793529-002

we can denote the interchange of the first and second input of a third-order tensor by $A_{jik}\mathbf{e}_i \otimes \mathbf{e}_j \otimes \mathbf{e}_k$. A tensor is symmetric with respect to the inputs i and j if the interchange does not alter the tensor, i. e., when $A_{ij} = A_{ji}$ in case of a second-order tensor. In case of fourth-order tensors, the superimposed T implies $\mathbb{C}^T = C_{klij}\mathbf{e}_i \otimes \mathbf{e}_j \otimes \mathbf{e}_k \otimes \mathbf{e}_l$. This is the mayor symmetry, if the tensor is symmetric w. r. t. interchanging the indices of the left- and right-index pairs, one speaks of the left and right minor symmetries or subsymmetries.

Antisymmetry implies that upon transposition the sign changes, e. g., $A_{ij} = -A_{ji}$ in case of a second-order tensor. A well-known instance is the permutation symbol

$$\varepsilon_{ijk} = \begin{cases} \{i,j,k\} = \{1,2,3\} \text{ or } \{2,3,1\} \text{ or } \{3,1,2\} & : 1 \\ \{i,j,k\} = \{3,2,1\} \text{ or } \{2,1,3\} \text{ or } \{1,3,2\} & : -1 \\ \text{otherwise} & : 0. \end{cases} \tag{2.5}$$

Amending ε_{ijk} with a basis gives the permutation tensor,

$$\overset{\langle 3\rangle}{\boldsymbol{\varepsilon}} = \varepsilon_{ijk}\mathbf{e}_i \otimes \mathbf{e}_j \otimes \mathbf{e}_k. \tag{2.6}$$

$\overset{\langle 3\rangle}{\boldsymbol{\varepsilon}}$ is antisymmetric in all inputs.

Placements

We distinguish the reference and current placement of a body. The index 0 indicates that the independent variable of a field is the location vector $\mathbf{x}_0$ in the reference placement. The same holds for the nabla operator: The index 0 indicates that the derivatives are w. r. t. x_{0i}. On other quantities, e. g., $\mathbb{C}_0$ in Chapter 13, the index 0 indicates that the quantity is a reference figure.

Gradients

The inputs that appear when taking a gradient are appended on the right. For example, the deformation gradient is defined as

$$\mathbf{F} = \mathbf{x}(\mathbf{x}_0) \otimes \nabla_0 = \partial x_i(x_{01}, x_{02}, x_{03})/\partial x_{0j}\mathbf{e}_i \otimes \mathbf{e}_j, \tag{2.7}$$

where the nabla operator is applied in the following way: $\cdot \times \nabla_{(0)} = (\partial \cdot /\partial x_{(0)i}) \times \mathbf{e}_i$. The derivative is taken before the product is carried out, which is important when curvilinear coordinates are used. Here, only the components are subjected to the derivative due to the use of Cartesian[1] coordinates, such that the tangential basis does not depend on the location.

1 René Descartes, 1596–1650 (https://en.wikipedia.org/wiki/Rene_Descartes).

2.2 Volume average

The volume average of a field $F(\mathbf{x})$ in a domain Ω is

$$\overline{F} = \frac{1}{V}\int_{\Omega} F(\mathbf{x})\,\mathrm{d}v = \langle F\rangle, \tag{2.8}$$

with the volume V of the domain Ω. The angular bracket is a common abbreviation. One needs to distinguish the volume average from the effective quantity, which we denote by a superscript $*$. The effective stiffness $\mathbb{C}^*$ is defined implicitly by

$$\overline{\boldsymbol{\sigma}} = \mathbb{C}^* : \overline{\boldsymbol{\varepsilon}} \tag{2.9}$$

and depends on the microstructure.

2.3 Voigt–Mandel notation

Occasionally, we use a normalized basis $\mathbf{E}_i$ for symmetric second-order tensors,

$$\mathbf{E}_1 = \mathbf{e}_1 \otimes \mathbf{e}_1, \quad \mathbf{E}_4 = \frac{1}{\sqrt{2}}(\mathbf{e}_1 \otimes \mathbf{e}_2 + \mathbf{e}_2 \otimes \mathbf{e}_1), \tag{2.10}$$

$$\mathbf{E}_2 = \mathbf{e}_2 \otimes \mathbf{e}_2, \quad \mathbf{E}_5 = \frac{1}{\sqrt{2}}(\mathbf{e}_1 \otimes \mathbf{e}_3 + \mathbf{e}_3 \otimes \mathbf{e}_1), \tag{2.11}$$

$$\mathbf{E}_3 = \mathbf{e}_3 \otimes \mathbf{e}_3, \quad \mathbf{E}_6 = \frac{1}{\sqrt{2}}(\mathbf{e}_2 \otimes \mathbf{e}_3 + \mathbf{e}_3 \otimes \mathbf{e}_2). \tag{2.12}$$

This is the modified Voigt[2] notation, also known as the Mandel[3] and Kelvin[4] notation. The basis is normalized such that

$$\mathbf{E}_i : \mathbf{E}_j = \delta_{ij}, \quad i,j = 1,\dots,6 \tag{2.13}$$

holds. It can be used to represent fourth-order tensors as 6×6-matrices. The normalization of the basis is important, since only then the usual rules of matrix calculus like the trace, determinant, matrix multiplication, Cramer's rule and many others can be applied directly on the 6×6 matrix components; see Brannon (2018), Sections 26.2 and 26.3, Bertram and Glüge (2015), Section 2.1.16, Helnwein (2001) or Cowin and Mehrabadi (1992). The normalization has been first used by Thomson (1856) (later 1. Baron Kelvin).

---

2 Woldemar Voigt senior, 1850–1913 (https://en.wikipedia.org/wiki/Woldemar_Voigt).

3 Jean Mandel, 1907–1982 (https://www.annales.org/archives/x/mandel.html).

4 William Thomson, 1. Baron Kelvin, 1824–1907 (https://en.wikipedia.org/wiki/William_Thomson,_1st_Baron_Kelvin).

2.4 Projectors

Projectors will prove to be invaluable tools. They are tensors with special properties. Projector systems are idempotent and orthogonal,

$$\overset{\langle 2n\rangle}{\mathbb{P}}_i \underbrace{\cdot\cdot\cdot\cdot\cdot}_{n\text{ scalar dots}} \overset{\langle 2n\rangle}{\mathbb{P}}_j = \begin{cases} \overset{\langle 2n\rangle}{\mathbb{P}}_i & \text{if } i = j \\ \overset{\langle 2n\rangle}{\mathbb{O}} & \text{if } i \neq j \end{cases} \tag{2.14}$$

and complete,

$$\sum_{i=1,k} \mathbb{P}_i = \mathbb{I}, \tag{2.15}$$

with $\mathbb{I}$ being the identity over the underlying vector space. Thus, projectors have only eigenvalues 0 and 1. Therefore, we also have

$$\overset{\langle 2n\rangle}{\mathbb{P}}_i \underbrace{\cdot\cdot\cdot\cdot\cdot}_{2n\text{ scalar dots}} \overset{\langle 2n\rangle}{\mathbb{P}}_j = \begin{cases} d_i & \text{with } d_i > 0 \text{ and integer if } i = j \\ 0 & \text{if } i \neq j \end{cases}. \tag{2.16}$$

d_i is the dimension of the subspace, into which the tensor of order n is projected to which the projector is applied. Commonly known projectors are $\mathbf{K} = (\mathbf{k}\cdot\mathbf{k})^{-1}\mathbf{k}\otimes\mathbf{k}$ and $\mathbf{I} - \mathbf{K}$, which project every vector into its part parallel and perpendicular to $\mathbf{k}$.

The most used projectors in material modeling are the eigenprojectors of diagonalizable second-order tensors

$$\mathbf{A} = \sum_{i=1,k} \lambda_k \mathbf{P}_k, \quad \mathbf{P}_k = \frac{1}{\mathbf{v}_k^r \cdot \mathbf{v}_k^l} \mathbf{v}_k^r \otimes \mathbf{v}_k^l \tag{2.17}$$

with the right and left eigenvectors $\mathbf{v}_k^{r/l}$. Other important projectors are the isotropic fourth-order projectors $\mathbb{P}_{\mathrm{I}1,2}$. Fourth-order tensors have several isotropic parts, but we usually need only the ones with major and subsymmetries. These isotropic parts are $\mathbf{I}\otimes\mathbf{I}$ and the identity $\mathbb{I}$ on symmetric second-order tensors. Any linear combination of these is isotropic as well. It is helpful to introduce an orthogonal basis, which leads to the projectors $\mathbb{P}_{\mathrm{I}1,2}$,

$$\mathbb{P}_{\mathrm{I}1} : \mathbf{A} = \frac{1}{3}\mathbf{I}\otimes\mathbf{I} : \mathbf{A} = \frac{\mathrm{tr}(\mathbf{A})}{3}\mathbf{I} =: \mathbf{A}^{\circ} \tag{2.18}$$

$$\mathbb{P}_{\mathrm{I}2} : \mathbf{A} = (\mathbb{I} - \mathbb{P}_{\mathrm{I}1}) : \mathbf{A} = \mathbf{A} - \mathbf{A}^{\circ} =: \mathbf{A}'. \tag{2.19}$$

They map every second-order tensor into their isotropic (dilatoric) and deviatoric part

$$\mathbf{A} = \mathbf{A}^{\circ} + \mathbf{A}'. \tag{2.20}$$

We can then use

$$\mathbb{C}_{\mathrm{Iso}} = 3K\mathbb{P}_{\mathrm{I}1} + 2G\mathbb{P}_{\mathrm{I}2}, \tag{2.21}$$

to construct isotropic tetrads, where $3K$ and $2G$ are the eigenvalues. The projection of an arbitrary tetrad $\mathbb{C}$ into its isotropic part is achieved with the help of the eighth-order projector

$$\overset{\langle 8\rangle}{\mathbb{P}} = \mathbb{P}_{I1} \otimes \mathbb{P}_{I1} + \frac{1}{\sqrt{5}}\mathbb{P}_{I2} \otimes \frac{1}{\sqrt{5}}\mathbb{P}_{I2}. \tag{2.22}$$

Note the normalization, which underlines the dual character of projectors: On one hand, in the spectral representation when they map second-order tensors to second-order tensors,

$$\mathbb{P}_{Ii} : \mathbb{P}_{Ij} = \delta_{ij}\mathbb{P}_{Ii}. \tag{2.23}$$

On the other hand, they can only be used as orthonormal bases for fourth-order tensors after normalization,

$$\mathbb{P}^*_{Ii} :: \mathbb{P}^*_{Ij} = \delta_{ij} \tag{2.24}$$

with

$$\mathbb{P}^*_{I1} = \mathbb{P}_{I1}, \tag{2.25}$$

$$\mathbb{P}^*_{I2} = \frac{1}{\sqrt{5}}\mathbb{P}_{I2}. \tag{2.26}$$

The numbers 1 and 5 correspond to the subspace dimensions of dilatoric and symmetric deviatoric second-order tensors. These have only 1 and 5 independent components. One obtains these values as the traces of the matrix components of $\mathbf{P}_{I1}$ and $\mathbf{P}_{I2}$ w. r. t. the Voigt–Mandel basis, since all projector eigenvalues are 0 or 1. With respect to this basis, the isotropic projectors have the following components:

$$\mathbb{P}_{I1} = \begin{bmatrix} \frac{1}{3} & \frac{1}{3} & \frac{1}{3} & & & \\ \frac{1}{3} & \frac{1}{3} & \frac{1}{3} & & & \\ \frac{1}{3} & \frac{1}{3} & \frac{1}{3} & & & \\ & & & & & \\ & & & & & \\ & & & & & \end{bmatrix} \mathbf{E}_i \otimes \mathbf{E}_j, \tag{2.27}$$

$$\mathbb{P}_{I2} = \mathbb{I} - \mathbb{P}_{I1} = \begin{bmatrix} \frac{2}{3} & -\frac{1}{3} & -\frac{1}{3} & & & \\ -\frac{1}{3} & \frac{2}{3} & -\frac{1}{3} & & & \\ -\frac{1}{3} & \frac{1}{3} & \frac{2}{3} & & & \\ & & & 1 & & \\ & & & & 1 & \\ & & & & & 1 \end{bmatrix} \mathbf{E}_i \otimes \mathbf{E}_j. \tag{2.28}$$

The traces are 1 and 5, respectively, the norms are 1 and $\sqrt{5}$.

2.5 Hooke's law

The Hookean[5] law presumes a linear stress-strain relation,

$$\boldsymbol{\sigma} = \mathbb{C} : \boldsymbol{\varepsilon}, \tag{2.29}$$

with $\boldsymbol{\sigma}$ and $\boldsymbol{\varepsilon}$ symmetric, which is why we can assume the left and right subsymmetries for $\mathbb{C}$, specifically $C_{ijkl} = C_{jikl} = C_{ijlk}$. Further, $\mathbb{C}$ has the major symmetry $C_{ijkl} = C_{klij}$, which is the integrability condition for the existence of an elastic energy[6]

$$w = \frac{1}{2}\boldsymbol{\varepsilon} : \mathbb{C} : \boldsymbol{\varepsilon} \tag{2.30}$$

with the potential relation

$$\boldsymbol{\sigma} = \frac{\partial w}{\partial \boldsymbol{\varepsilon}} = \frac{1}{2}\mathbb{C} : \boldsymbol{\varepsilon} + \frac{1}{2}\boldsymbol{\varepsilon} : \mathbb{C} = \frac{1}{2}(\mathbb{C} + \mathbb{C}^T) : \boldsymbol{\varepsilon}. \tag{2.31}$$

One can see that the antisymmetric part of $\mathbb{C}$ vanishes upon differentiation. From a thermodynamic point of view, elasticity laws that cannot be written as derivatives of elastic energies can be discarded, since otherwise energy conservation is violated. The elasticity law $\boldsymbol{\sigma}(\boldsymbol{\varepsilon})$ needs to be integrable to $w(\boldsymbol{\varepsilon})$.[7] Strains are dimensionless, which is why the physical unit of stiffness is equal to the physical unit of stress, $\mathrm{N/m^2}$ and at the same time the physical unit of energy per volume $\mathrm{Nm/m^3}$.

Inverting $\mathbb{C}$ on the subspace of tetrads with right and left subsymmetries gives the compliance tetrad $\mathbb{S} = \mathbb{C}^{-1}$ for the inverted Hookean law,

$$\boldsymbol{\varepsilon} = \mathbb{S} : \boldsymbol{\sigma}. \tag{2.32}$$

$\mathbb{S}$ has the same symmetries as $\mathbb{C}$ does. The complementary energy is

$$w^* = \frac{1}{2}\boldsymbol{\sigma} : \mathbb{S} : \boldsymbol{\sigma}, \tag{2.33}$$

with the potential relation

$$\boldsymbol{\varepsilon} = \frac{\partial w^*}{\partial \boldsymbol{\sigma}}. \tag{2.34}$$

In linear elasticity, $w(\boldsymbol{\varepsilon}) = w^*(\boldsymbol{\sigma}(\boldsymbol{\varepsilon}))$. In nonlinear elasticity, this does not hold, then w and w^* are their mutual Legendre[8] transforms.

5 Robert Hooke, 1635–1703 (https://en.wikipedia.org/wiki/Robert_Hooke).

6 George Green, 1793–1841 (https://en.wikipedia.org/wiki/George_Green_(mathematician)).

7 "Integrability" refers to specific symmetries in multidimensional analysis. In real analysis, these symmetries do not appear. Then, "integrability" refers to diverging integrals due to poles.

8 Adrien-Marie Legendre, 1752–1833 (https://en.wikipedia.org/wiki/Adrien-Marie_Legendre).

2.6 The isotropic Hookean law

In the case of isotropy, we have with equation (2.21),

$$\boldsymbol{\sigma} = 3K\boldsymbol{\varepsilon}^{\circ} + 2G\boldsymbol{\varepsilon}' \tag{2.35}$$
$$= 3K\boldsymbol{\varepsilon}^{\circ} + 2G(\boldsymbol{\varepsilon} - \boldsymbol{\varepsilon}^{\circ}) \tag{2.36}$$
$$= (3K - 2G)\boldsymbol{\varepsilon}^{\circ} + 2G\boldsymbol{\varepsilon} \tag{2.37}$$
$$= \underbrace{(K - 2G/3)}_{\lambda}(\boldsymbol{\varepsilon} : \mathbf{I})\mathbf{I} + 2\mu\boldsymbol{\varepsilon}. \tag{2.38}$$

The last line is Lamé's[9] representation with the coefficients $\lambda = K - 2G/3$ and $\mu = G$. The inversion is obtained by using the dilator-deviator decomposition (equations (2.18) to (2.20))

$$\boldsymbol{\sigma}^{\circ} = 3K\boldsymbol{\varepsilon}^{\circ} \quad \leftrightarrow \quad \boldsymbol{\varepsilon}^{\circ} = (3K)^{-1}\boldsymbol{\sigma}^{\circ}, \tag{2.39}$$
$$\boldsymbol{\sigma}' = 2GK\boldsymbol{\varepsilon}' \quad \leftrightarrow \quad \boldsymbol{\varepsilon}' = (2G)^{-1}\boldsymbol{\sigma}'. \tag{2.40}$$

This is inverted easily w. r. t. the projector representation,

$$\boldsymbol{\varepsilon} = \left((3K)^{-1}\mathbb{P}_{I1} + (2G)^{-1}\mathbb{P}_{I2}\right) : \boldsymbol{\sigma}. \tag{2.41}$$

One just needs to take the reciprocals of the eigenvalues while maintaining the eigenprojectors.

The anisotropic Hookean law is discussed in Section 13.3.

2.7 Generalization of real functions to tensors

From the projector properties (equation (2.14)), it follows that powers of diagonalizable tensors can be obtained by simply raising the eigenvalues to that power in the spectral representation. For a second-order tensor with three different eigenvalues, this is

$$\underbrace{\mathbf{A} \cdot \mathbf{A} \cdot \cdots \cdot \mathbf{A}}_{n \text{ factors}} = \mathbf{A}^n = (\lambda_1\mathbf{P}_1 + \lambda_2\mathbf{P}_2 + \lambda_3\mathbf{P}_3)^n \tag{2.42}$$
$$= \lambda_1^n\mathbf{P}_1 + \lambda_2^n\mathbf{P}_2 + \lambda_3^n\mathbf{P}_3, \tag{2.43}$$

since all mixed projector products are zero and the unmixed projector products give just the projector (idempotence of the projectors). Thus we can generalize any real function that can be written as a power series to second-order tensors by applying the said function to the eigenvalues in the spectral representation. One has, e. g., for the logarithm of a tensor $\mathbf{A}$

$$\ln \mathbf{A} = \ln(\lambda_1)\mathbf{P}_1 + \ln(\lambda_2)\mathbf{P}_2 + \ln(\lambda_3)\mathbf{P}_3. \tag{2.44}$$

---

9 Gabriel Lamé, 1795–1870 (https://en.wikipedia.org/wiki/Gabriel_Lam%C3%A9).

For eigenvalues with a multiplicity, the eigenvalue is factored out, and the bracket is the corresponding eigenprojector, just as in the case of $\mathbb{C}_{\mathrm{Iso}}$ in equation (2.21) the eigenvalue $2G$ has five linear independent eigentensors.

2.8 List of symbols

Symbol	Meaning
Ω	domain of a representative volume element or material sample
$\partial\Omega$	boundary of Ω
$\chi(\mathbf{x})$	indicator function
A	area
$p_{ij\ldots k}$	n-point correlation function
v_i	volume fraction of a phase, $0 \le v_i \le 1$
V	volume of a domain Ω
$\mathbf{d}$	difference vector between two locations $\mathbf{d} = \mathbf{x}_1 - \mathbf{x}_2$
$\mathbf{f}$	force vector $\mathbf{f} = A\mathbf{t}$
$\mathbf{g}$	temperature gradient $\mathbf{g} = \nabla T$
$\mathbf{k}$	wave vector in the Fourier series $\mathbf{f}(\mathbf{x}) = \sum_{\forall \mathbf{k}} e^{i\mathbf{x}\cdot\mathbf{k}} \hat{\mathbf{f}}_{\mathbf{k}}$
$\mathbf{n}_{(0)}$	normal vector
$\mathbf{p}$	polarization field
$\mathbf{q}$	heat flux vector
$\mathbf{t}$	stress vector
$\mathbf{u}$	displacement vector $\mathbf{u} = \mathbf{x} - \mathbf{x}_0$
$\mathbf{x}$	location vector in the current placement
$\mathbf{x}_0$	location vector in the reference placement
$\mathbf{I}$	second-order identity tensor
$\boldsymbol{\varepsilon}$	linear strain tensor $\boldsymbol{\varepsilon} = \mathrm{sym}(\mathbf{H})$
$\mathbf{F}$	deformation gradient $\mathbf{F} = \mathbf{x}(\mathbf{x}_0) \otimes \nabla_0$
$\mathbf{H}$	displacement gradient $\mathbf{H} = \mathbf{u}(\mathbf{x}_0) \otimes \nabla_0$
$\mathbf{L}$	constitutive tensor for vector field problems, e. g., $\mathbf{q} = \mathbf{L} \cdot \mathbf{g}$
$\mathbf{P}$	context dependent the first Piola–Kirchhoff stresses, a projection tensor or the polarization field
$\boldsymbol{\sigma}$	Cauchy's stress tensor
$\boldsymbol{\tau}$	polarization stress $\boldsymbol{\tau}(\mathbf{x}) = (\mathbb{C}(\mathbf{x}) - \mathbb{C}^0) : \boldsymbol{\varepsilon}(\mathbf{x})$
$\mathbb{I}$	identity on symmetric second-order tensors, $\mathrm{sym}(\mathbf{A}) = \mathbb{I} : \mathbf{A}$ $\mathbb{I} = \frac{1}{2}(\delta_{ik}\delta_{jl} + \delta_{il}\delta_{jk})\mathbf{e}_i \otimes \mathbf{e}_j \otimes \mathbf{e}_k \otimes \mathbf{e}_l$
$\overset{\langle 3 \rangle}{\boldsymbol{\varepsilon}}$	third-order permutation tensor
$\mathbb{C}$	stiffness tetrad
$\mathbb{C}^0$	comparison stiffness in the difference problem
$\mathbb{C}_{\#}$	anisotropic reference stiffness
$\mathbb{K}$	concentration tensor, maps via $\boldsymbol{\sigma}(\mathbf{x}) = \mathbb{K}(\mathbf{x})\overline{\boldsymbol{\sigma}}$ mean stresses to local stresses
$\mathbb{L}$	concentration tensor, maps via $\boldsymbol{\varepsilon}(\mathbf{x}) = \mathbb{L}(\mathbf{x})\overline{\boldsymbol{\varepsilon}}$ mean strains to local strains
$\mathbb{P}_{C1,2,3}$	the three cubic projectors
$\mathbb{P}_{I1/2}$	first and second isotropic tensors of fourth order
∇	nabla operator $\nabla(\cdot) = \partial(\cdot)/\partial x_{(0)i}\mathbf{e}_i$
∇^2	Laplace operator $\nabla^2(\cdot) = ((\cdot) \otimes \nabla) \cdot \nabla = (\cdot)_{,ii}$
Δ	difference between two quantities of the same kind, e. g., $\Delta\mathbb{C} = \mathbb{C} - \mathbb{C}^0$

3 Characterization of microstructure

We first establish geometric and statistical descriptions of microstructures.

3.1 Indicator function

For a specific microstructure, we use the indicator function $\chi_i(\mathbf{x})$ to indicate whether or not we find material i at the point $\mathbf{x}$. $\chi_i(\mathbf{x})$ takes the value 0 when at $\mathbf{x}$ the material i is not found, and correspondingly the value 1 when material i is found at $\mathbf{x}$. This implies the assumption of a piecewise homogeneous phase distribution; see Figure 3.1. Assuming that we have n phases, we can conclude from the absence of $n-1$ phases the presence of the remaining phase. Therefore, we can express one indicator function by all other indicator functions:

$$\chi_n(\mathbf{x}) = 1 - \chi_1(\mathbf{x}) - \chi_2(\mathbf{x}) \cdots - \chi_{n-1}(\mathbf{x}). \tag{3.1}$$

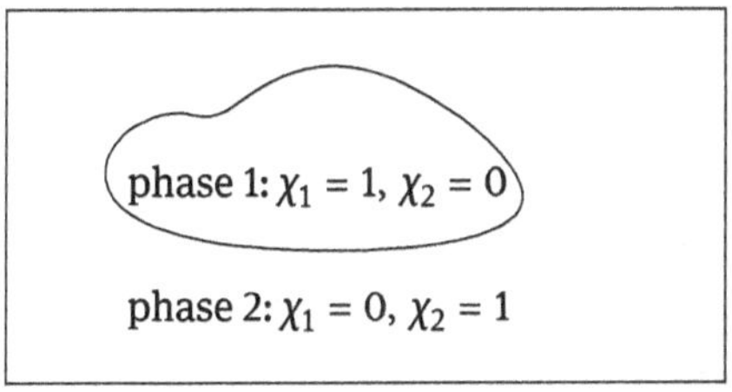

Figure 3.1: The indicator function indicates which phase is found at point **x**.

3.2 One point statistic

By $p_i(\mathbf{x})$, we denote the probability to find material i at a random (uniformly distributed) point $\mathbf{x}$. It is

$$1 = \sum_{i=1}^{n} p_i(\mathbf{x}), \tag{3.2}$$

where n is the number of phases. p_i is termed as a one-point statistic or one-point correlation, as only one point is involved. It corresponds to the volume fraction

$$v_i = p_i = \frac{1}{V} \int_{\Omega} \chi_i(\mathbf{x})\, \mathrm{d}v. \tag{3.3}$$

The integral is over a domain Ω with volume V, which is large enough to be called representative or statistically sufficient. We assume that the microstructure is statis-

https://doi.org/10.1515/9783110793529-003

tically homogeneous, which means that the location of the domain window Ω is not important for the statistical properties.

Since we can write one indicator function in terms of all other indicator functions, we can also express one correlation function by the other ones,

$$v_n = p_n = \frac{1}{V} \int_{\Omega} 1 - \chi_1(\mathbf{x}) - \chi_2(\mathbf{x}) \cdots - \chi_{n-1}(\mathbf{x}) \, \mathrm{d}v \tag{3.4}$$

$$= 1 - p_1 - p_2 \cdots - p_{n-1}. \tag{3.5}$$

3.3 Multipoint statistics

The two-point correlation function

$$p_{ii}(\mathbf{x}_1, \mathbf{x}_2), \tag{3.6}$$

gives the probability of finding phase i simultaneously at the points $\mathbf{x}_1$ and $\mathbf{x}_2$. Due to the presumed statistical homogeneity, we can subtract $\mathbf{x}_2$ from both arguments, such that

$$p_{ii}(\mathbf{x}_1, \mathbf{x}_2) = p_i(\underbrace{\mathbf{x}_1 - \mathbf{x}_2}_{\mathbf{d}}, \mathbf{o}) \tag{3.7}$$

with the difference vector $\mathbf{d}$. We can therefore reduce the number of arguments of multipoint correlations by one. The two-point correlation is

$$p_{ii}(\mathbf{d}) = \frac{1}{V} \int_{\Omega} \chi_i(\mathbf{x}) \chi_i(\mathbf{x} + \mathbf{d}) \, \mathrm{d}v. \tag{3.8}$$

It depends only on $\mathbf{d}$ and can therefore be visualized easily. Let us consider two examples.

Example 1

Consider a laminate of a 0.4 mm (phase 1) and a 0.6 mm (phase 2) layer that is repeated periodically (Figure 3.2). What are the 1-, 2- and 3-point correlations of this structure?

Solution

Let x be the coordinate in normal direction. At $x = 0$, a 0.4 mm thick layer of phase 1 begins. We drop the unit of length in the remainder. With this, we can set up the indicator functions:

$$\chi_1(x) = \begin{cases} 0 \leq \operatorname{mod}(x, 1) < 0.4 : 1, \\ 0.4 \leq \operatorname{mod}(x, 1) < 1 : 0, \end{cases} \tag{3.9}$$

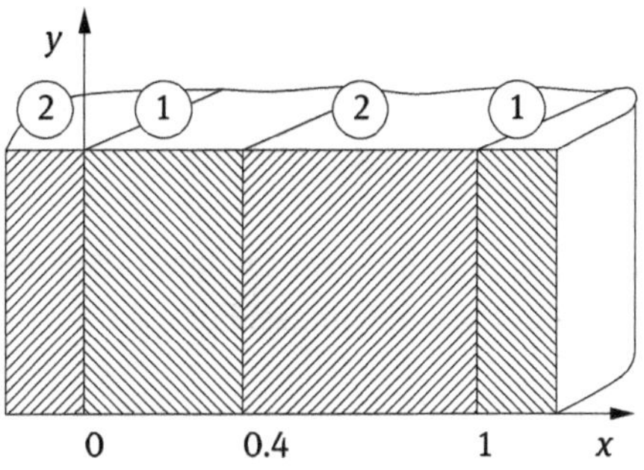

Figure 3.2: Periodic laminate with $p_1 = 0.4$ and $p_2 = 0.6$.

$$\chi_2(x) = \begin{cases} 0 \leq \mathrm{mod}(x,1) < 0.4 : 0, \\ 0.4 \leq \mathrm{mod}(x,1) < 1 : 1. \end{cases} \tag{3.10}$$

This gives the volume fractions:

$$v_1 = \int_0^1 \chi_1(x)\,\mathrm{d}x = \int_0^{0.4} \mathrm{d}x = 0.4, \tag{3.11}$$

$$v_2 = \int_0^1 \chi_2(x)\,\mathrm{d}x = \int_{0.4}^{1.0} \mathrm{d}x = 1 - 0.4 = 0.6. \tag{3.12}$$

The integration is carried out over the interval $0 \leq x \leq 1$, which is representative of the microstructure. Beyond the unit interval that repeats, the indicator functions are periodic. This periodicity is inherited to the correlation functions. The first statistical information regarding the phases' arrangement is contained in $p_{11}(d)$ with $d = x_1 - x_2$,

$$p_{11}(d) = \int_0^1 \chi_1(x)\chi_1(x+d)\,\mathrm{d}x \tag{3.13}$$

$$= \int_0^{0.4} \chi_1(x+d)\,\mathrm{d}x. \tag{3.14}$$

Rewriting the second indicator function is not as simple as before due to the dependence on d. When $0 < d < 0.4$, we have $\chi_1(x+d) = 1$ in the interval $0 < x < 0.4 - d$. When $0.4 < d < 0.6$, there is no x for which $\chi_1(x+d) = 1$. When $0.6 < d < 1.0$, we have $\chi_1(x+d) = 1$ in the interval $1 - d < x < 0.4$. The periodicity is transferred to d. This finally gives

$$p_{11}(d) = \begin{cases} 0.0 \le \mathrm{mod}(d,1) < 0.4 : \int_0^{0.4-d} \mathrm{d}x = 0.4 - d \\ 0.4 \le \mathrm{mod}(d,1) < 0.6 : 0 \\ 0.6 \le \mathrm{mod}(d,1) < 1 : \int_{1-d}^{0.4} \mathrm{d}x = d - 0.6 \end{cases} \tag{3.15}$$

This can be comprehended easily by laying a line of length d over the laminate and checking for which starting points x both ends lie in the same phase. For $p_{22}(d)$, one obtains upon using $\chi_1(x) + \chi_2(x) = 1$,

$$p_{22}(d) = \int_0^1 \chi_2(x)\chi_2(x+d)\,\mathrm{d}x \tag{3.16}$$

$$= \int_0^1 (1-\chi_1(x))(1-\chi_1(x+d))\,\mathrm{d}x \tag{3.17}$$

$$= \int_0^1 1 - \chi_1(x) - \chi_1(x+d) + \chi_1(x)\chi_1(x+d)\,\mathrm{d}x \tag{3.18}$$

$$= 1.0 - 0.4 - \int_0^1 \chi_1(x+d)\,\mathrm{d}x + p_{11}(d) \tag{3.19}$$

$$= 0.6 - \int_{0+d}^{1+d} \chi_1(x)\,\mathrm{d}x + p_{11}(d) \tag{3.20}$$

$$= 0.6 - 0.4 + p_{11}(d) \tag{3.21}$$

$$= 0.2 + p_{11}(d) \tag{3.22}$$

$$= \begin{cases} 0 \le \mathrm{mod}(d,1) < 0.4 : 0.6 - d \\ 0.4 \le \mathrm{mod}(d,1) < 0.6 : 0.2 \\ 0.6 \le \mathrm{mod}(d,1) < 1 : d - 0.4. \end{cases} \tag{3.23}$$

Further, we have the mixed correlations $p_{12}(d) = p_{21}(d)$. The index symmetry is a consequence of the presumed statistical homogeneity:

$$p_{12}(d) = \int_0^1 \chi_1(x)\chi_2(x+d)\,\mathrm{d}x \tag{3.24}$$

$$= \int_0^1 (1-\chi_2(x))(1-\chi_1(x+d))\,\mathrm{d}x \tag{3.25}$$

$$= \int_0^1 1 - \chi_2(x) - \chi_1(x+d) + \chi_2(x)\chi_1(x+d)\,\mathrm{d}x. \tag{3.26}$$

The input $x + d$ in the third summand can be transferred to the integration bounds. In our example, this has no effect. For nonperiodic structures, this has no effect if the domain is sufficiently large:

$$p_{12}(d) = \int_0^1 1\,\mathrm{d}x - \int_0^1 \chi_2(x)\,\mathrm{d}x - \int_{0+d}^{1+d} \chi_1(x) + \int_0^1 \chi_2(x)\chi_1(x+d)\,\mathrm{d}x \quad \leftarrow \quad \text{statistical homogeneity} \tag{3.27}$$

$$p_{12}(d) = \int_0^1 \underbrace{1 - \chi_2(x) - \chi_1(x)}_{0} + \chi_2(x)\chi_1(x+d)\,\mathrm{d}x \tag{3.28}$$

$$= p_{21}(d). \tag{3.29}$$

With the sum of probabilities $1 = p_{11}(d) + p_{22}(d) + 2p_{12}(d)$, we obtain

$$p_{12}(d) = \frac{1}{2}(1 - p_{11}(d) - p_{22}(d)), \tag{3.30}$$

therefore,

$$p_{12}(d) = \begin{cases} 0 \leq \mathrm{mod}(d,1) < 0.4 : d \\ 0.4 \leq \mathrm{mod}(d,1) < 0.6 : 0.4 \\ 0.6 \leq \mathrm{mod}(d,1) < 1 : 1 - d. \end{cases} \tag{3.31}$$

The two-point correlations are depicted in Figure 3.3. The three-point correlations can be reduced to two parameters d_1 and d_2. One recognizes the periodic structure and the symmetry when $d_1 = d_2$; see Figure 3.4.

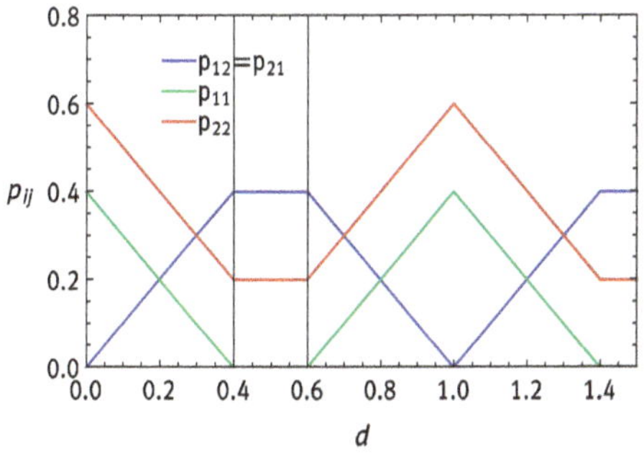

Figure 3.3: Two-point correlations $p_{ij}(d)$ of the laminate. One recognizes the periodicity of the structure. Further, one can see that p_{11} is zero between $d = 0.4$ and $d = 0.6$, which allows to conclude the layer thicknesses. The thicknesses of 0.4 mm and 0.6 mm are marked as vertical lines.

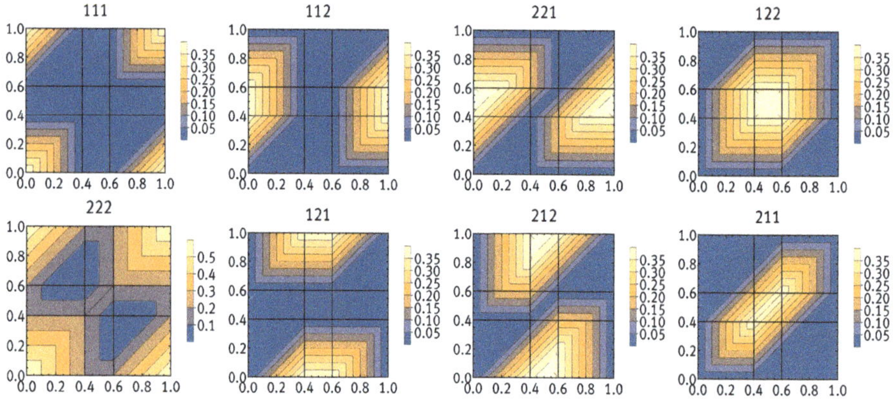

Figure 3.4: Three-point correlations $p_{ijk}(d_1, d_2) = \int_\Omega \chi_i(x)\chi_j(x + d_1)\chi_k(x + d_2)\,\mathrm{d}x$ with d_1 horizontally and d_2 vertically, the critical lengths 0.4 mm and 0.6 mm are marked. The results can be repeated periodically. Swapping the indices j and k results in a mirroring along the diagonal $d_1 = d_2$. For $j = k$, the result is symmetric w. r. t. the diagonal $d_1 = d_2$.

Example 2

We examine the microstructure given in Figure 3.5 and estimate the anisotropy using the two-point correlation.

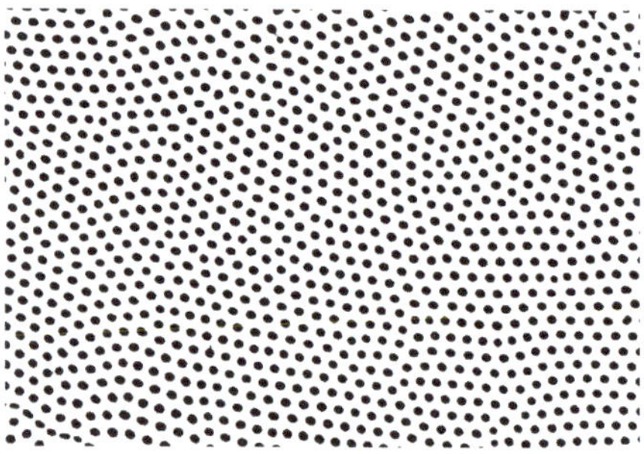

Figure 3.5: A real matrix-inclusion structure (from Orera and Merino, 2015).

The pixel colors black and white correspond to the matrix indicator function, taking the values 0 in the black phase and 1 in the white phase. The evaluation the defining integrals becomes a summation over all pixels. We evaluate the two-point correlation for a large number of difference vectors **d**; the values are then plotted in the **d**-plane. This is implemented in the computer math system, Mathematica. The code is given in Listing 3.1. The resulting plots show a preferred direction for the short-range order; see Figure 3.6.

Listing 3.1: Mathematica script (https://gitlab.com/gluegerainer/listings-homogenization-methods/-/blob/main/Listing-01_en.nb) for generating the two-point correlations on a binary microstructure.

```
(* The numerical evaluation of the two-point correlation is quicker when the function is
   compiled. *)
(* grid[[i,j]] corresponds to the value of the indicator function, replace by (1-grid[[i,j]])
   to invert. *)
averageCompiled = Compile[{{x, _Integer}, {y, _Integer}, {grid, _Real, 2}},
   Module[{dimx, dimy}, {dimx, dimy} = Dimensions[grid];
    Mean@Flatten@
      Table[grid[[i, j]] grid[[i + x - 1, j + y - 1]] (* this is the product of the two
    indicator functions *),
        {i, dimx - x + 1}, {j, dimy - y + 1}]], CompilationTarget -> "C", RuntimeOptions -> "
    Speed"];

(* Read the microstructure and binarize the image. *)
TempDat = ImageData[Binarize[Import["bild2_binaer.jpeg"]]];

(* Save two-point correlation (tpcf) as a grid and show. *)
tpcf = ParallelTable[N[averageCompiled[i, j, TempDat]], {i, 1, 100, 1}, {j, 1, 100, 1}];
ListContourPlot[tpcf, PlotRange -> All, Mesh -> 20,
 MeshFunctions -> {Function[{x, y, z}, z]},
 Contours -> {{0.8, {Thick, Black}}, ## & @@ ({#, None} & /@ Range[Min@tpcf, Max@tpcf, .0025])},
 PlotLegends -> Placed[BarLegend[Automatic, LegendMargins -> {{0, 0}, {10, 5}}], Right]]
```

3.4 Periodic microstructures

For periodic microstructures, we have

$$p_i(\mathbf{x}) = p_i(\mathbf{x} + k\mathbf{p}_1 + l\mathbf{p}_2 + m\mathbf{p}_3) \tag{3.32}$$

for all integers k, l, m. The three linearly independent vectors $\mathbf{p}_i$ span the periodicity frame. The correlations behave as in the first example: they are periodic and do not approach an asymptotic value for $\|\mathbf{d}\| \to \infty$. Periodic microstructures have a long-range order. In the real-world example above, one recognizes the fading short-range order in Figure 3.6. In a strict sense, the microstructure is not periodic. But if a small domain is considered, one may be able to speak of a periodic component in the short-range order.

3.5 Other statistical descriptors

The above correlations allow for a complete statistical description of the microstructure. They appear in the integrals for the effective properties in analytic homogenization; see, e. g., Chapter 15 in Milton (2002).

Nevertheless, a variety of statistical descriptors have been proposed (Torquato, 2005) which are tailored to specific aspects of a microstructure. For example, the lineal path function gives the probability for a needle of length l and direction $\mathbf{v}$ to lie

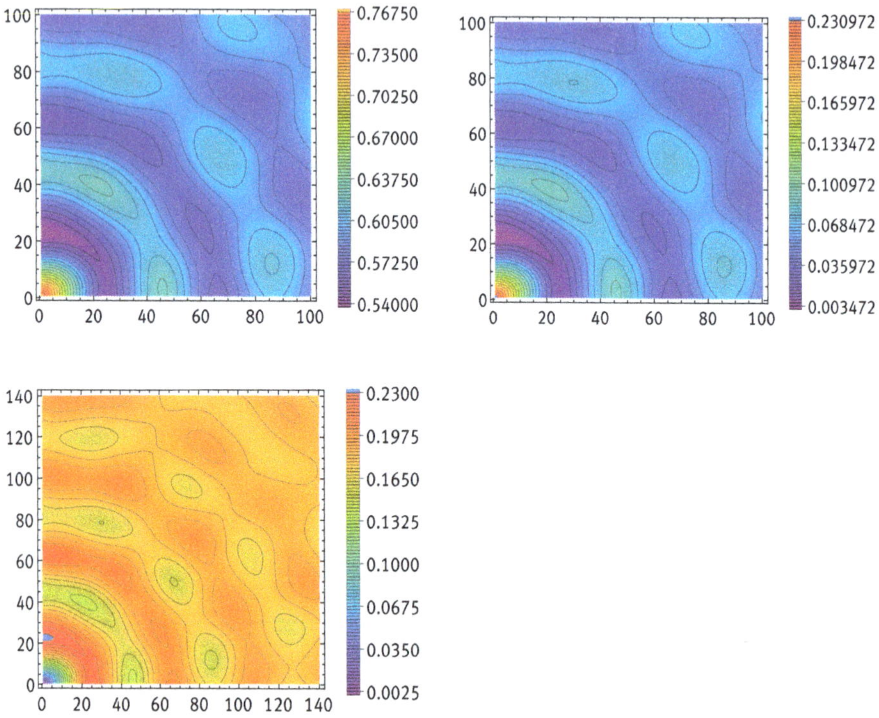

Figure 3.6: p_{11} (left top), p_{22} (right top) and p_{12} (bottom) for the real microstructure in Figure 3.5. At $\mathbf{d} = \mathbf{o}$ one recovers the volume fractions phases 0.768483 (for p_{11}) and 0.230972 (for p_{22}) and the value 0 (for p_{12}). We recognize regular rings with distances of approximately 45 pixels, which is more or less the average distance of the inclusions. This is the short-range order. With increasing distance, the noise takes over, and the two-point correlation approaches the expectation value of two randomly selected points $0.768483^2 = 0.590567$ (p_{11}) and $0.230972^2 = 0.05335$ (p_{22}). Thus there is no long-range order. Further, one sees a weak anisotropy in the first ring, the maxima of which correspond to the most frequent location difference vectors between two particles.

completely in one phase. Contrary to the two-point correlation function, the lineal path function decreases monotonically with l.

3.6 Generating material distributions for given correlation functions

Sometimes the inverse problem arises: Correlations are given, e. g., from measurements, and a representative material sample with these properties is to be generated numerically. There are no direct solution methods for this. Instead, microstructures are generated by trial and error and adopted to fit the given correlations. An introduction to this topic can be found in the section "Heterogeneous media reconstruction" in Tashkinov (2017).

3.7 Summary

A complete description of a microstructure with piecewise homogeneous phases is given by indicator functions $\chi_i(\mathbf{x})$. The statistical properties are derived from these as correlation functions. They quantify and objectify a microstructure, such that these can be compared and analyzed. The correlation functions allow quantifying the structural anisotropy, short- and long-range order and periodicity.

Unfortunately, even simple microstructures allow rarely for deriving closed expressions for the correlations. Commonly, the correlations are determined numerically from micrographs or 3D tomographic data.

Further, the correlations appear in integrals for the effective properties when analytic methods are applied entirely in real space; see Chapter 15 in Milton (2002) or Chapter 20 in Torquato (2005). However, this is mathematically very complicated. In Fourier space, however, the Fourier transforms of the indicator function of a repeatable unit cell take the place of the correlation functions, which is much less complicated; see Section 12.

Exercise: Needle problem

What is the probability of a needle of length l that is randomly dropped on a Cartesian grid of spacing l to not cross a grid line (Figure 3.7)?

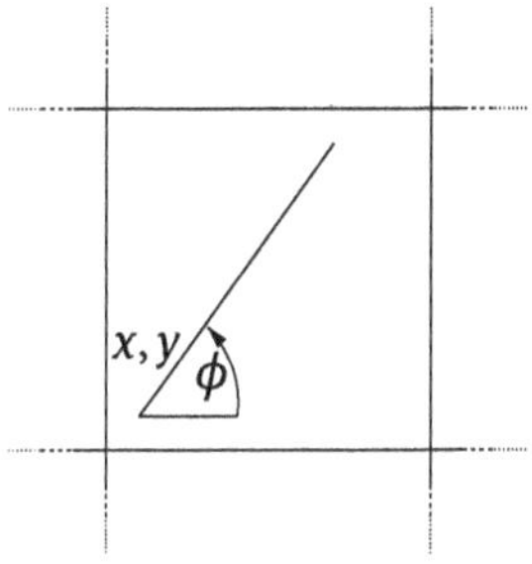

Figure 3.7: Needle on a grid.

Solution

$(\pi-3)/\pi \approx 0.045 = 4.5\,\%$. The state space of the needle is parameterized by one end lying at the location $x = 0\dots l$ and $y = 0\dots l$ w. r. t. a grid point and the angle $\phi = 0\dots 2\pi$ with the x-axis. One needs to construct the function $\chi(x,y,\phi)$, the value of which is 0 if the needle crosses a grid line and 1 otherwise. Integration over this function over the state space divided by the state space size is the fraction of realizations without needle-grid intersections:

$$\chi(x,y,\phi) = \begin{cases} x + l\cos\phi < l \quad \text{and} \quad y + l\cos\phi < l : 1 \\ \text{else} : 0 \end{cases} \tag{3.33}$$

The integration is reduced to the part of the state space with $\chi = 1$, using the symmetry w. r. t. the four corners of the grid square,

$$\int_0^{2\pi}\int_0^{l}\int_0^{l} \chi(x,y,\phi)\,\mathrm{d}x\,\mathrm{d}y\,\mathrm{d}\phi = 4\int_0^{\pi/2}\int_0^{l-l\sin\phi}\int_0^{l-l\cos\phi} \mathrm{d}x\,\mathrm{d}y\,\mathrm{d}\phi \tag{3.34}$$

$$= 2l^2(\pi - 3). \tag{3.35}$$

Integrating over the space of all realizations gives $2\pi l^2$. The fraction of realizations without needle-grid intersections is

$$\frac{2l^2(\pi-3)}{2\pi l^2} = (\pi - 3)/\pi. \tag{3.36}$$

This value can be approximated easily by numerical experiments; see Listing 3.2.

Listing 3.2: Numerical experiment (https://gitlab.com/gluegerainer/listings-homogenization-methods/-/blob/main/Listing-02_en.nb) to approach the probability of the needle not intersecting the grid lines. Example outputs are `0.04513355` and `0.0450703414486`.

```
n = 20000000;
Print["Approximation = ", ParallelSum[
    phi = Random[Real, {0, 2 Pi}]; (*random angle*)
    x = Random[];(*x and y between 0 and 1*)
    y = Random[];
    If[Floor[x + Cos[phi]] == 0 && Floor[y + Sin[phi]] == 0, 1, 0], {i, 1, n}]/n // N,
    " vs. exact = ", (Pi - 3)/Pi // N]
```

4 The initial and boundary value problem of homogenization

Virtual material sample

We consider a virtual material sample as depicted in Figure 4.1. The domain is abbreviated as Ω, the boundary by $\partial\Omega$. The displacement field $\mathbf{u}(\mathbf{x}_0)$ contains the movement or deviation of the sample from the initial position, for which $\mathbf{u}$ is zero everywhere. $\mathbf{x}_0$ is the location vector in the initial position,

$$\mathbf{x}(\mathbf{x}_0) = \mathbf{u}(\mathbf{x}_0) + \mathbf{x}_0 \tag{4.1}$$

is the location vector in the deformed position. The displacement gradient is

$$\mathbf{H} = \mathbf{u}(\mathbf{x}_0) \otimes \nabla_0. \tag{4.2}$$

The stress vector field is defined on the boundary $\partial\Omega$ and is written as

$$\mathbf{t}(\mathbf{x}_0) = \mathbf{P}(\mathbf{x}_0) \cdot \mathbf{n}_0(\mathbf{x}_0), \tag{4.3}$$

with the first Piola[1]–Kirchhoff[2] stress tensor $\mathbf{P}$ and the normal vector $\mathbf{n}_0(\mathbf{x}_0)$ in the initial placement. All of these quantities are fields. We use mostly the location vector $\mathbf{x}_0$ as the independent variable. Often this dependency is not denoted explicitly but implied.

Small strains

In the case of small strains and small rotations, the strain tensor is

$$\boldsymbol{\varepsilon} = \frac{1}{2}(\mathbf{H} + \mathbf{H}^T), \tag{4.4}$$

i. e., the symmetric part of $\mathbf{H}$. In the following derivation, the transition to small strains is obtained by replacing the first Piola–Kirchhoff stresses $\mathbf{P}$ with the Cauchy[3] stresses $\boldsymbol{\sigma}$ ($\mathbf{P} \to \boldsymbol{\sigma}$) and by $\mathbf{H} \to \boldsymbol{\varepsilon}$, $\mathbf{x}_0 \to \mathbf{x}$ and $\nabla_0 \to \nabla$. The following holds for large strains, for which the derivation is, without the symmetrization, more compact than for small strains.

1 Gabrio Piola, 1794–1850 (https://en.wikipedia.org/wiki/Gabrio_Piola).

2 Gustav Robert Kirchhoff, 1824–1887 (https://en.wikipedia.org/wiki/Gustav_Robert_Kirchhoff).

3 Augustin-Louis Cauchy, 1789–1857 (https://en.wikipedia.org/wiki/Augustin-Louis_Cauchy).

https://doi.org/10.1515/9783110793529-004

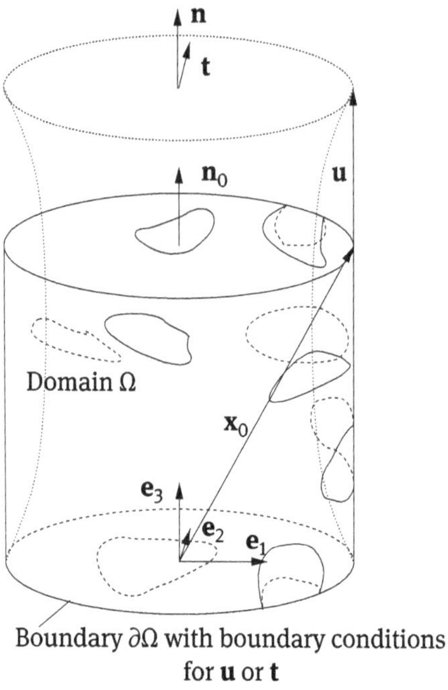

Figure 4.1: Virtual material sample with microstructure.

Boundary conditions

In the most general case, the boundary conditions are linear constraints between $\mathbf{u}$ and $\mathbf{t}$ in three orthogonal directions. For example, in case of a frictionless, stiff surface with normal $\mathbf{n}_0$ we have $\mathbf{u} \cdot \mathbf{n}_0 = 0$ and $\mathbf{t} \cdot (\mathbf{I} - \mathbf{n} \otimes \mathbf{n}) = \mathbf{o}$. In case of an elastic foundation with stiffness $\mathbf{C}$ we have $\mathbf{t} = \mathbf{C} \cdot \mathbf{u}$. Mostly, either $\mathbf{t}$ or $\mathbf{u}$ is prescribed directly. Prescribing $\mathbf{u}$ one speaks of Dirichlet[4] or direct boundary conditions, since the value of the unknown function $\mathbf{u}$ is prescribed directly. Prescribing $\mathbf{t}$ we speak of Neumann[5] boundary conditions and when $\mathbf{t} = \mathbf{o}$ of natural boundary conditions. For example, all finite element codes presume stress-free boundaries $\mathbf{t} = \mathbf{o}$ if nothing else is specified.

Representative volume element

R. Hill[6] coined the term "representative volume element" for a virtual material sample. The expression seems to imply that the volume element needs to tessellate the entire space. This is by no means necessary, which is why the term "virtual material

4 Peter Gustav Lejeune Dirichlet, 1805–1859 (https://en.wikipedia.org/wiki/Peter_Gustav_Lejeune_Dirichlet).

5 Carl Gottfried Neumann, 1832–1925 (https://en.wikipedia.org/wiki/Carl_Gottfried_Neumann).

6 Rodney Hill, 1911–2011 (https://en.wikipedia.org/wiki/Rodney_Hill).

sample" is more precise. *The term "representative" refers to the sample size alone.* The sample should be large enough in all spatial directions to cover the significant statistical properties; see Section 4.2. Then one can measure the effective Young's[7] modulus on cylindrical samples, as usual. Nevertheless, the term "RVE" is well established as a concept, which is why we use it here. Further, a periodic continuability of the RVE is suited very well to simulate the embedding of the material sample in an identical material, which is why periodic boundary conditions are commonly used to approximate the effective properties already at relatively small RVE sizes; see Section 4.3.5.

4.1 The differential equation

The initial and boundary value problem described above is completed by a differential equation in the domain Ω. It consists of a material law $\mathbf{P}(\mathbf{H},\ldots)$ and the equilibrium conditions. Depending on the material model, internal variables (e. g., plastic strains) or a rate dependence may appear. The material law depends on the location $\mathbf{x}_0$ and can be written formally as a sum of products of individual laws with the corresponding indicator function,

$$\mathbf{P}(\mathbf{x}_0) = \sum_{i=1\ldots n} \chi_i(\mathbf{x}_0)\mathbf{P}_i(\mathbf{H}(\mathbf{x}_0),\ldots). \tag{4.5}$$

The equilibrium conditions are

$$\mathbf{P}\cdot\nabla_0 = \mathbf{o}. \tag{4.6}$$

In homogenization, we neglect body forces $\rho\mathbf{b}$ and inertia forces $\rho\mathbf{a}$, since these are not material properties that are subjected to homogenization. Hence the right side of the last equation is $\mathbf{o}$[8]

The governing DE is a system of three partial differential equations of degree two (two derivatives in $\mathbf{x}_0$). The right side is always zero, i. e., the PDE is homogeneous. The PDE coefficients on the left are by definition of an inhomogeneous domain not constant since the material law depends on the location $\mathbf{x}_0$.

In this direct form, the problem is solved usually only numerically; see Section 4.4. For analytical methods, this representation is inadequate, especially the nonconstant coefficients (equation (4.5)) are problematic. Therefore, in the simplest setting of small strain linear elasticity, the problem is reformulated as the eigenstrain problem and the polarization problem; see Section 8.

---

7 Thomas Young, 1773–1829 (https://en.wikipedia.org/wiki/Thomas_Young_(scientist)).

8 The sometimes encountered arguments that the RVE volume is proportional to l^3_{mini} (see Section 4.2) and the surface is proportional to l^2_{mini}, which leads with $l_{\text{mini}} \ll l_{\text{macro}}$ to the conclusion that body forces are small compared to contact forces is invalid. The effective material properties are obtained for $l_{\text{mini}} \to \infty$. Therefore, body forces do not vanish automatically; they need to be excluded explicitly.

4.2 Scale separation

A large distance between the micro and macroscale is needed to apply effective material laws. Well-known examples of such effective properties from physics are, e. g., entropy and temperature, which are only defined for sufficiently complex systems, but not for few or even individual particles. For large scale separations, the effective material behavior approaches the expected value, similar to the law of large numbers.

According to Hashin's[9] micro-mini-macro principle (Hashin, 1983), the scale separation needs to bridge even two scales. On one hand, the material sample should be representative and, therefore, larger than the inhomogeneities on the microscale, but it replaces the material behavior on a point-like location on the macroscale,

$$l_{\text{micro}} \ll l_{\text{mini}} \ll l_{\text{macro}}. \tag{4.7}$$

l_{micro} is the characteristic size of the inhomogeneities on the microscale, l_{mini} is the size of a representative material sample and l_{macro} is the approximate size of a part; see Figure 4.2.

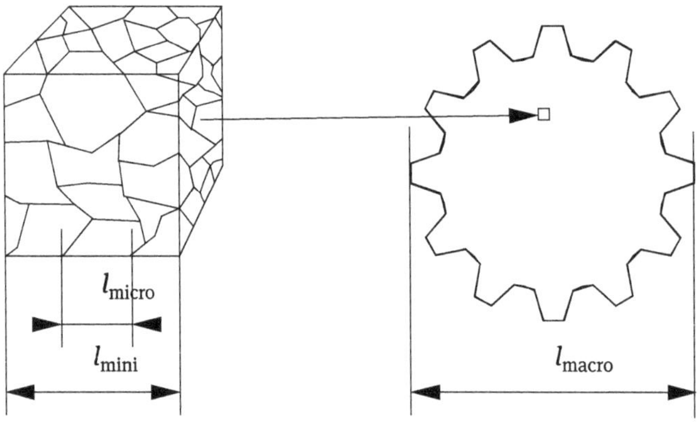

Figure 4.2: Scale separation for a grain structure.

Example

The effect of an insufficient scale separation can be demonstrated experimentally. The following example is for a block of polypropylene, produced by injection molding. Due to faster cooling of the outer layers, the sample has a casting skin; see Figure 4.3. To determine the local material properties, a tensile sample is sliced into layers of 50 μm thickness. The slices are subjected to tensile tests, and Young's modulus is measured for each slice. Young's modulus is measured as well for the bulk sample.

9 Zvi Hashin, 1929–2017 (https://en.wikipedia.org/wiki/Zvi_Hashin).

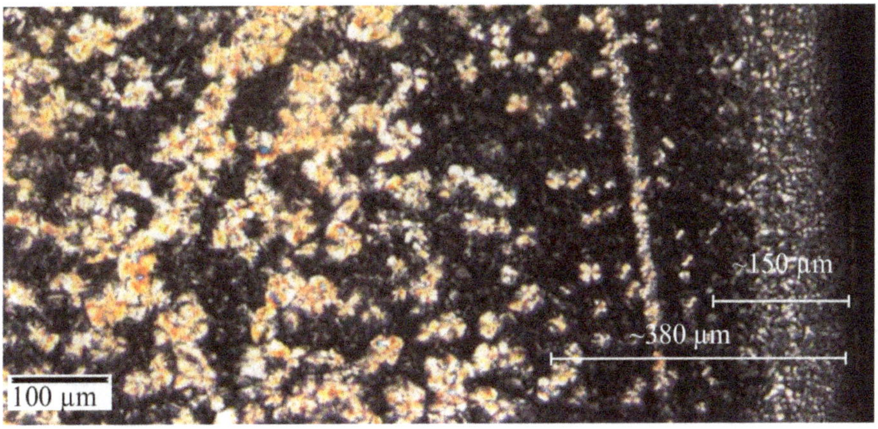

Figure 4.3: The microstructure of an injection molded sample of polypropylene depends on the distance from the surface, which cools down much quicker than the bulk of the sample. On the surface, an amorphous polymeric phase is predominant, while in the slower cooling regions, spherulitic grains have enough time to develop.

Young's modulus measured for the slices depends on the surface distance; see the blue curve in Figure 4.4. The mean value is around 920 MPa, which is much smaller than Young's modulus for the entire sample of around 1320 MPa. Since one may think of the slices as parallel springs, one may expect their average Young modulus to equal the bulk Young modulus. But this is not the case.

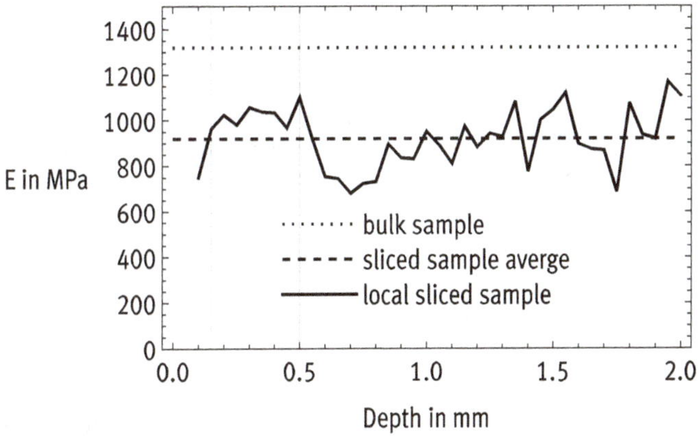

Figure 4.4: Young's modulus of the slice as a function of the distance from the surface. The average is approximately 920 MPa (dashed line). Young's modulus of the bulk sample is approximately 1320 MPa (dotted line).

The cause for this discrepancy is the violation of scale separation. The slices have practically everywhere a plane stress state due to the closeness of the free surface to any material point. This makes no difference for homogeneous materials, which would show a homogeneous, uniaxial tensile stress state, independent of the sample thickness, and one would measure always the same Young's modulus. But due to the inhomogeneities, there are local stress fluctuations. These can spread freely as 3D stress states in the bulk sample but are confined to plane stress states in the slices. Hence, the slices lack a direction for load redistribution, which is why they appear softer than the bulk; see Figure 4.5 (Glüge, 2020). One can check the micrograph in Figure 4.3 that the inhomogeneities, which are spherulitic grains in the bulk, are of the size of the slice thickness of about 50 μm; hence there is no scale separation along the thickness direction.

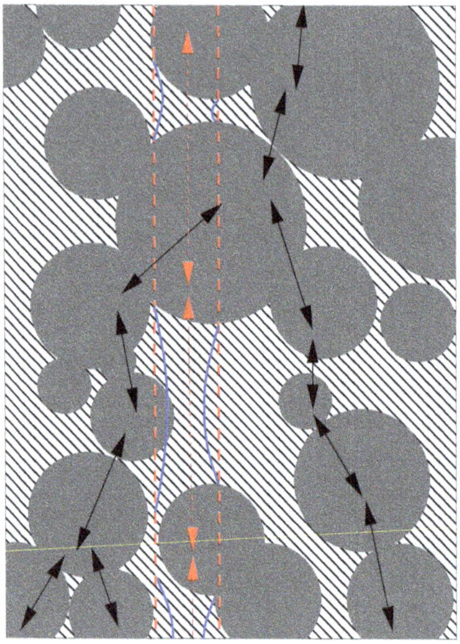

Figure 4.5: A microstructure with stiff inclusions (semicrystalline spherulites) in a soft matrix (amorphous polymeric phase). The loading path is concentrated along the inclusions (black arrows). In a slice (red), the loading path is forced into the softer phase. Moreover, the lateral straining can adjust freely in the thin sample (blue). Due to these two effects, the thin sample is softer.

4.3 Integral theorems of homogenization

We are often interested in volume averages. These can be rewritten often as surface integrals by using the theorem of Gauß[10]–Ostrogradski.[11] The surface integrals can sometimes be evaluated by the boundary conditions alone, which does not require the solution of the boundary value problem.

10 Carl Friedrich Gauß, 1777–1855 (https://en.wikipedia.org/wiki/Carl_Friedrich_Gauss).

11 Michail Wassiljewitsch Ostrogradski, 1801–1861 (https://en.wikipedia.org/wiki/Mikhail_Ostrogradsky).

4.3.1 Effective strains

The mean value of the displacement gradient is

$$\overline{\mathbf{H}} = \frac{1}{V}\int_{\Omega} \mathbf{u} \otimes \nabla_0 \, dv_0 \tag{4.8}$$

$$= \frac{1}{V}\int_{\partial\Omega} \mathbf{u} \otimes \mathbf{n}_0 \, dv_0. \tag{4.9}$$

For $\overline{\boldsymbol{\varepsilon}}$, this needs to be merely symmetrized. Hence we can integrate the effective strains from the boundary displacements alone. This holds for any boundary value problem in continuum mechanics.

Homogeneous strain boundary conditions

Fixing the displacement gradient $\mathbf{H}_\partial$ and applying the boundary displacements on the entire surface by

$$\mathbf{u}_\partial(\mathbf{x}_0) = \mathbf{H}_\partial \cdot \mathbf{x}_0 \tag{4.10}$$

yields

$$\overline{\mathbf{H}} = \frac{1}{V}\int_{\partial\Omega} \mathbf{H}_\partial \cdot \mathbf{x}_0 \otimes \mathbf{n}_0 \, dv_0 \tag{4.11}$$

$$= \mathbf{H}_\partial \cdot \underbrace{\frac{1}{V}\int_{\partial\Omega} \mathbf{x}_0 \otimes \mathbf{n}_0 \, dv_0}_{\mathbf{I}} = \mathbf{H}_\partial; \tag{4.12}$$

hence the effective strains can be prescribed by such boundary conditions.

4.3.2 Effective stresses

A similar relation exists for the stresses, but only if body forces are absent. First, we rewrite the volume integral over the Piola–Kirchhoff stresses $\mathbf{P}$ by using the product rule backward,

$$\overline{\mathbf{P}}^T = \frac{1}{V}\int_{\Omega} \mathbf{P}^T(\mathbf{x}_0) \, dv_0 \tag{4.13}$$

$$= \frac{1}{V}\int_{\Omega} [(\mathbf{x}_0 \otimes \mathbf{P}(\mathbf{x}_0)) \cdot \nabla_0 - \mathbf{x}_0 \otimes (\mathbf{P}(\mathbf{x}_0) \cdot \nabla_0)] \, dv_0. \tag{4.14}$$

This expansion is not obvious. It is easily seen when using indices (we do not write out the dependence on $\mathbf{x}_0$ and the index 0) w. r. t. the basis $\mathbf{e}_i \otimes \mathbf{e}_j$,

$$\frac{1}{V}\int_{\Omega} P_{ji}\, dv_0 = \frac{1}{V}\int_{\Omega} \left[(x_i P_{jk})_{,k} - x_i P_{jk,k}\right] dv_0 \tag{4.15}$$

$$= \frac{1}{V}\int_{\Omega} x_{i,k} P_{jk} + x_i P_{jk,k} - x_i P_{jk,k}\, dv_0 \tag{4.16}$$

$$= \frac{1}{V}\int_{\Omega} \delta_{ik} P_{jk}\, dv_0 \tag{4.17}$$

$$= \frac{1}{V}\int_{\Omega} P_{ji}\, dv_0. \tag{4.18}$$

With the local balance equation without body forces $\mathbf{P}(\mathbf{x}_0) \cdot \nabla_0 = \mathbf{o}$ the second integral in equation (4.14) vanishes. The first integral is recast as a surface integral by the Gauß–Ostrogradski theorem:

$$\overline{\mathbf{P}}^T = \frac{1}{V}\int_{\Omega} (\mathbf{x}_0 \otimes \mathbf{P}(\mathbf{x}_0)) \cdot \nabla_0\, dv_0 \tag{4.19}$$

$$= \frac{1}{V}\int_{\partial\Omega} \mathbf{x}_0 \otimes \mathbf{P}(\mathbf{x}_0) \cdot \mathbf{n}_0\, dv_0, \tag{4.20}$$

or with $\mathbf{t} = \mathbf{P} \cdot \mathbf{n}_0$ and without the transposition

$$\overline{\mathbf{P}} = \frac{1}{V}\int_{\partial\Omega} \mathbf{t} \otimes \mathbf{x}_0\, dv_0. \tag{4.21}$$

For the Cauchy stresses and small strains $\overline{\boldsymbol{\sigma}}$ it is merely necessary to symmetrize. To conclude, we can integrate the mean stresses $\overline{\boldsymbol{\sigma}}$ from the stresses at the boundary $\mathbf{t}(\mathbf{x}_0)$. This generally holds for static problems without body forces.

Homogeneous stress boundary conditions

By fixing a stress tensor $\mathbf{P}_\partial$ and prescribing the stress vector at each boundary point by

$$\mathbf{t}_\partial(\mathbf{x}_0) = \mathbf{P}_\partial \cdot \mathbf{n}_0(\mathbf{x}_0) \tag{4.22}$$

we prescribe the effective stresses automatically,

$$\overline{\mathbf{P}} = \frac{1}{V}\int_{\Omega} \mathbf{P}_\partial \cdot \mathbf{n}_0 \otimes \mathbf{x}_0\, dv_0 \tag{4.23}$$

$$= \mathbf{P}_\partial \cdot \underbrace{\frac{1}{V}\int_{\Omega} \mathbf{n}_0 \otimes \mathbf{x}_0\, dv_0}_{\mathbf{I}} = \mathbf{P}_\partial. \tag{4.24}$$

4.3.3 The Hill–Mandel condition

The stress power (Hill, 1963; Mandel, 1966) is an elementary quantity in continuum mechanics. It plays an important role in energy conservation. It is, therefore, reasonable to require the conservation of stress power when changing scales,

$$\langle \mathbf{P}(\mathbf{x}) \rangle : \langle \dot{\mathbf{H}}(\mathbf{x}) \rangle = \langle \mathbf{P}(\mathbf{x}) : \dot{\mathbf{H}}(\mathbf{x}) \rangle. \tag{4.25}$$

On the left-hand side, the effective stress power is obtained from the effective stresses and the effective strain rate, i. e., first, averaging, then taking the scalar product. On the right-hand side, the stress power is calculated locally, and then the mean value is taken. One might say that we require ergodicity when averaging the stress power. Since we denote everything w. r. t. the location in the reference placement $\mathbf{x}_0$, the material time derivative and the nabla operator ∇_0 can be taken in arbitrary order.

The question arises of which restriction arises from equation (4.25).

- Initially, it was thought that the above equation is only satisfied for infinitely large material samples since only then the statistical fluctuation average out. Therefore, the Hill–Mandel condition was thought of as a quality indicator for the size of the material sample.
- One can show that the Hill–Mandel condition is satisfied for specific boundary conditions a priori, independent of the material sample size and the material behavior. Therefore, today the Hill–Mandel condition is considered a restriction on the valid boundary conditions. For homogenization, only boundary conditions are valid that satisfy the Hill–Mandel condition a priori.

It is easy to see that with a decomposition of the displacement gradient

$$\dot{\mathbf{H}}(\mathbf{x}_0) = \underbrace{\dot{\overline{\mathbf{H}}}}_{\langle \dot{\mathbf{H}} \rangle} + \underbrace{\dot{\tilde{\mathbf{H}}}(\mathbf{x}_0)}_{\dot{\mathbf{H}}(\mathbf{x}_0) - \langle \dot{\mathbf{H}} \rangle} \tag{4.26}$$

into its homogeneous part $\dot{\overline{\mathbf{H}}}$ and its fluctuating part $\dot{\tilde{\mathbf{H}}}(\mathbf{x}_0)$ the Hill–Mandel condition can be rewritten as

$$0 = \langle \mathbf{P}(\mathbf{x}_0) : \dot{\tilde{\mathbf{H}}}(\mathbf{x}_0) \rangle. \tag{4.27}$$

The decomposition of $\mathbf{H}$ induces a decomposition of the displacement field into a linear and a fluctuating part:

$$\mathbf{u}(\mathbf{x}_0) = \overline{\mathbf{H}} \cdot \mathbf{x}_0 + \tilde{\mathbf{u}}(\mathbf{x}_0). \tag{4.28}$$

With the product rule backwards (equation (4.30) to equation (4.31)), the local balance $\mathbf{P}(\mathbf{x}_0) \cdot \nabla_0 = \mathbf{o}$ (equation (4.31) to equation (4.32)), the theorem of Gauß–Ostrogradski (equation (4.31) to equation (4.32)) and Cauchy's theorem $\mathbf{t} = \mathbf{P} \cdot \mathbf{n}_0$ (equation (4.32) to equation (4.33)), equation (4.27) becomes

$$0 = \int_{\partial\Omega} \dot{\tilde{\mathbf{u}}}(\mathbf{x}_0) \cdot \mathbf{t}(\mathbf{x}_0)\, da_0. \tag{4.29}$$

The derivation is best denoted in indices starting from equation (4.27),

$$0 = \int_{\Omega} P_{ij}(\mathbf{x}_0)\dot{\tilde{u}}_{i,j}(\mathbf{x}_0)\, dv_0 \tag{4.30}$$

$$0 = \int_{\Omega} (P_{ij}(\mathbf{x}_0)\dot{\tilde{u}}_i(\mathbf{x}_0))_{,j} - \underbrace{P_{ij,j}(\mathbf{x}_0)}_{0}\, \dot{\tilde{u}}_i(\mathbf{x}_0)\, dv_0 \tag{4.31}$$

$$0 = \int_{\partial\Omega} P_{ij}(\mathbf{x}_0)\dot{\tilde{u}}_i(\mathbf{x}_0)\, n_{0j}\, da_0 \tag{4.32}$$

$$0 = \int_{\partial\Omega} t_i(\mathbf{x}_0)\dot{\tilde{u}}_i(\mathbf{x}_0)\, da_0 \tag{4.33}$$

$$0 = \int_{\partial\Omega} \mathbf{t}(\mathbf{x}_0) \cdot \dot{\tilde{\mathbf{u}}}(\mathbf{x}_0)\, da_0. \tag{4.34}$$

In this form, the Hill–Mandel condition can be easily checked for given boundary conditions $\mathbf{t}$ and $\mathbf{u}$.

Homogeneous strain boundary conditions

To check the Hill–Mandel condition for homogeneous strain boundary conditions, the form equation (4.34) is suited best. On the boundary, we have $\dot{\mathbf{u}}(\mathbf{x}_0) = \dot{\mathbf{H}}_\partial \cdot \mathbf{x}_0$. With $\dot{\overline{\mathbf{H}}} = \dot{\mathbf{H}}_\partial$ from equation (4.12), it is clear that $\dot{\tilde{\mathbf{u}}}(\mathbf{x}_0)$ is zero, therefore, the Hill–Mandel condition is satisfied.

Homogeneous stress boundary conditions

Equation (4.29) becomes with the boundary condition $\mathbf{t}(\mathbf{x}_0) = \mathbf{P}_\partial \cdot \mathbf{n}_0(\mathbf{x}_0)$,

$$0 = \int_{\partial\Omega} \dot{\tilde{\mathbf{u}}}(\mathbf{x}_0) \otimes \mathbf{n}_0(\mathbf{x}_0)\, da_0 : \mathbf{P}_\partial. \tag{4.35}$$

The integral is $\int_{\partial\Omega} \dot{\tilde{\mathbf{H}}}(\mathbf{x}_0)\, dv_0$, which vanishes by definition.

4.3.4 Change of the independent variable in case of homogeneous stress boundary conditions

Usually, we write our (effective) material laws with the strains as the independent variable and the stresses as the dependent variable. Therefore, it is somewhat impractical to prescribe the stresses and observe the strains of our virtual material sample. We, therefore, change now to the strains $\overline{\mathbf{H}}$ as the independent variable, but maintain the homogeneous stress on the boundary.

To do so, we need the following argument from variational calculus. For the true motion of the body, the global stress power[12]

$$\dot{W}(\dot{\mathbf{u}}(\mathbf{x}_0)) = \int_{\partial\Omega} \mathbf{t}(\mathbf{x}_0) \cdot \dot{\mathbf{u}}(\mathbf{x}_0)\, da_0 \tag{4.36}$$

takes a minimum with respect to all admissible velocity fields. $\dot{W}$ is a functional of the velocity field. In linear elasticity without body and inertia forces, this becomes simply the principle of the minimum of the potential energy. We decompose the stresses $\mathbf{P}$ into a homogeneous and a fluctuating part, or likewise $\mathbf{t}(\mathbf{x}_0)$ into a linear and a fluctuating part,

$$\mathbf{P}(\mathbf{x}_0) = \overline{\mathbf{P}} + \widetilde{\mathbf{P}}(\mathbf{x}_0), \tag{4.37}$$

$$\mathbf{t}(\mathbf{x}_0) = \overline{\mathbf{P}} \cdot \mathbf{n}_0 + \widetilde{\mathbf{P}}(\mathbf{x}_0) \cdot \mathbf{n}_0, \tag{4.38}$$

and insert this into the global stress power:

$$\dot{W}(\dot{\mathbf{u}}(\mathbf{x}_0)) = \int_{\partial\Omega} (\overline{\mathbf{P}} \cdot \mathbf{n}_0 + \widetilde{\mathbf{P}}(\mathbf{x}_0) \cdot \mathbf{n}_0) \cdot \dot{\mathbf{u}}(\mathbf{x}_0)\, da_0. \tag{4.39}$$

We now consider the first variation w. r. t. $\dot{\mathbf{u}}(\mathbf{x}_0)$. The variation $\delta\dot{\mathbf{u}}(\mathbf{x}_0)$ has to be compatible with the boundary conditions. Our only constraint is that we want to prescribe effective strains,

$$\dot{\overline{\mathbf{H}}} = \frac{1}{V} \int_{\partial\Omega} \dot{\mathbf{u}}(\mathbf{x}_0) \otimes \mathbf{n}_0(\mathbf{x}_0)\, da_0. \tag{4.40}$$

Hence the variation should not affect the average strains, i. e., $\delta\dot{\mathbf{u}}(\mathbf{x}_0)$ must satisfy the equation

12 By doing this, we confine ourselves to so-called "Generalized Standard Materials," for which a dissipation potential and a free energy need to be specified (Halphen and Son Nguyen, 1975). These comply automatically with the first and second laws of thermodynamics. The GSM framework is very general and covers all relevant materials, such that this restriction is not of practical relevance.

$$\delta\dot{\overline{\mathbf{H}}} = \mathbf{O} = \frac{1}{V}\int_{\partial\Omega} \delta\dot{\mathbf{u}}(\mathbf{x}_0) \otimes \mathbf{n}_0(\mathbf{x}_0)\, da_0. \tag{4.41}$$

The first variation of $\dot{W}$ w. r. t. $\delta\dot{\mathbf{u}}(\mathbf{x}_0)$ vanishes,

$$\delta\dot{W} = 0 = \int_{\partial\Omega} (\overline{\mathbf{P}} \cdot \mathbf{n}_0 + \tilde{\mathbf{t}}(\mathbf{x}_0)) \cdot \delta\dot{\mathbf{u}}(\mathbf{x}_0)\, da_0. \tag{4.42}$$

The first summand is zero because of equation (4.41). It remains

$$0 = \int_{\partial\Omega} \tilde{\mathbf{t}}(\mathbf{x}_0) \cdot \delta\dot{\mathbf{u}}(\mathbf{x}_0)\, da_0, \tag{4.43}$$

what needs to hold for arbitrary variations $\delta\dot{\mathbf{u}}(\mathbf{x}_0)$ that are compatible with equation (4.41). This is only possible for $\tilde{\mathbf{t}}(\mathbf{x}_0) = \mathbf{o}$, because the choice $\tilde{\mathbf{t}}(\mathbf{x}_0) = \alpha\delta\dot{\mathbf{u}}$ with $\alpha > 0$ is always admissible, since it does not violate the decomposition equation (4.37). This results in a positive integrant and, finally, in a nonzero first variation.

Therefore, homogeneous stress boundary conditions are realized simply by prescribing only the effective deformations, equation (4.40). Although these are identical to the homogeneous stress boundary conditions, one speaks of *kinematic minimal boundary conditions* when $\dot{\overline{\mathbf{H}}}$ is the independent variable.

4.3.5 Generalized boundary conditions that satisfy the Hill–Mandel condition

The observation that the homogeneous boundary conditions satisfy the Hill–Mandel condition raises the question whether one can give the entire class of boundary conditions that conform to the Hill–Mandel condition. This generalization is

- The boundary $\partial\Omega$ is divided into parts $\partial_i\Omega$, which do not need to be simply connected.
- An average deformation $\dot{\overline{\mathbf{H}}}$ is prescribed.
- On each part $\partial_i\Omega$ kinematic minimal boundary conditions are prescribed:

$$\underbrace{\dot{\overline{\mathbf{H}}} \cdot \int_{\partial\Omega_i} \mathbf{x}_0 \otimes \mathbf{n}_0\, dv_0}_{\dot{\overline{\mathbf{H}}}_i} = \int_{\partial\Omega_i} \dot{\mathbf{u}}(\mathbf{x}_0) \otimes \mathbf{n}_0(\mathbf{x}_0)\, dv_0. \tag{4.44}$$

We can identify the most important special cases. The extremal cases, namely a maximal partitioning of $\partial\Omega$ and no partitioning of $\partial\Omega$ correspond to the homogeneous boundary conditions.

- The *homogeneous strain boundary conditions* correspond to an infinitely fine partitioning, such that $\dot{\mathbf{u}}(\mathbf{x}_0) = \dot{\overline{\mathbf{H}}} \cdot \mathbf{x}_0$ is imposed pointwise.

- The *periodic boundary conditions* are obtained when the partial surfaces $\partial\Omega_i$ are pairwise contracted at points with opposing surface normals $\mathbf{n}_{0+} = -\mathbf{n}_{0-}$,

$$\dot{\bar{\mathbf{H}}} \cdot (\mathbf{x}_{0+} - \mathbf{x}_{0-}) = \dot{\mathbf{u}}(\mathbf{x}_{0+}) - \dot{\mathbf{u}}(\mathbf{x}_{0-}). \tag{4.45}$$

  The periodicity is a tool to select pairs of points with opposing surface normals, but other pointwise couplings with opposing surface normals are possible without contradicting the above equations. Obviously, then the periodicity is lost, as in case of the homogeneous boundary conditions. This is not necessarily a drawback, as the periodicity implies an artificial anisotropy.
- Without a partitioning of $\partial\Omega$, the left integral becomes $V\mathbf{I}$. One is left with the kinematic minimal boundary conditions, which are equivalent to the homogeneous stress boundary conditions.

A sketch of the different boundary conditions is given in Figure 4.6.

Example: Boundary conditions in a tensile test
In a tensile test, a cylindrical specimen is subjected to elongation by displacing the top surfaces while leaving the lateral surface traction-free. Are these boundary conditions compatible with the Hill–Mandel condition? Yes, they are. The lateral surface $\partial\Omega_{\text{lateral}}$ is one surface part with homogeneous stress boundary conditions with $\mathbf{t} = 0$; the top surfaces are infinitely fine partitioned, such that homogeneous strain boundary conditions are applied, with the resulting homogeneous displacement of the surface in the normal direction. Not surprisingly, effective properties can be determined in tensile tests when the samples are large enough.

One can see that an additional constraint equation is imposed on every partition of the boundary, and each additional constraint results in reaction stresses, which manifest as higher effective stresses. While the boundary can freely deform under kinematic minimal boundary conditions (i. e., homogeneous stress boundary conditions), it is subjected to an affine deformation in case of homogeneous strain boundary conditions. In the first case, we have a much smaller kinematic constraint and, therefore, lower effective stresses. In the second case, we have much more constraints with corresponding reaction stresses, and hence higher effective stresses.

Theoretically, this is not an issue: Increasing the characteristic size l of the material sample, its boundary grows proportional to l^2, while the volume grows proportional to l^3. Therefore, we expect to observe convergence to the effective material properties in any case as we increase the size of the material sample, independent of the boundary conditions. Unfortunately, we cannot calculate or experiment with infinitely large samples. Therefore, it is reasonable to choose boundary conditions with a moderate amount of kinematic constraint (one speaks of the stiffness of boundary conditions). These mimic the embedding of the sample in a matrix with a similar material behavior, as if the sample was infinite in size. By this, one tries to start as close

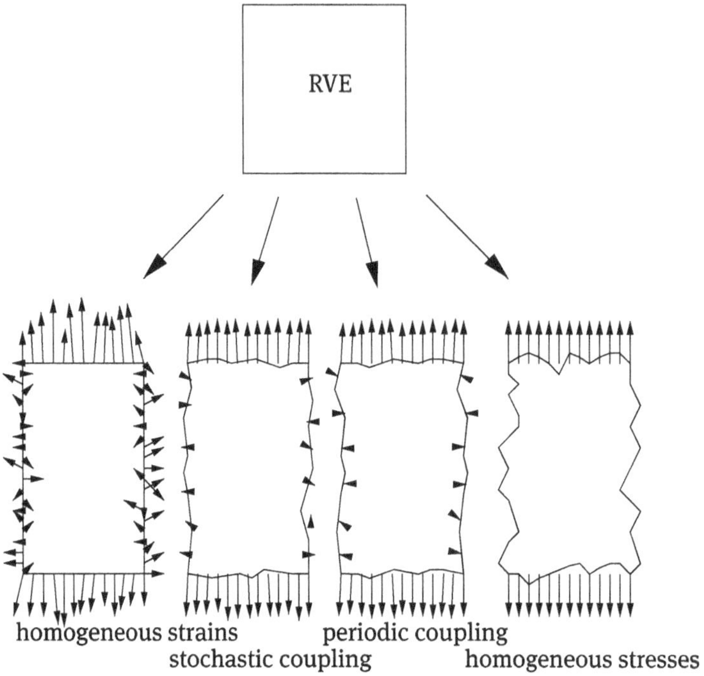

Figure 4.6: Different RVE boundary conditions.

as possible to the limit $l \to \infty$. In this regard, the periodic boundary conditions are established as the standard boundary conditions for RVE, probably partly due to the misleading term "RVE" instead of the virtual material sample. They are a reasonable compromise between the homogeneous boundary conditions, but they induce a artificial periodicity frame that does not exist in stochastic microstructures. Although this periodicity frame vanishes when $l \to \infty$, it can be avoided entirely by randomly coupling the surface points summarized in one $\partial\Omega_i$. Further, the amount of kinematic constraint can be freely adjusted when the number of surface points that are coupled needs not to be two.

It turns out that the regularity of periodic boundary conditions provides a frame for RVE-wide plastic localizations in the form of shear bands, which can be suppressed by stochastic boundary conditions. Therefore, the stiffness and localization resistance of the boundary conditions can be adjusted independently by stochastic or otherwise nonperiodic coupling (Glüge, 2013).

4.4 Numerical homogenization

Numerical homogenization is often referred to as the RVE method. One solves the RVE boundary value problem numerically, usually with the aid of the finite element method. The effective deformation is prescribed by appropriate boundary conditions (mostly periodic boundary conditions), and the resulting effective, average stresses are extracted. This then gives discrete values of the effective stresses-effective strain-relation for a specific loading path. This method is applicable for arbitrarily complicated microstructures and intricate material laws. Contrary to analytic homogenization methods, no closed-form expression for the effective material law is obtained.

Necessary RVE size

As a criterion for convergence to the effective material behavior upon increasing the RVE size, one compares two differently sized RVEs, the effective material responses of which need to be sufficiently close to each other. One finds that convergence depends strongly on the material behavior under consideration. For anisotropic, elastic homogenization tasks, the required RVE size increases approximately linearly with the degree of anisotropy and inverse quadratically with the acceptable error (Nygards, 2003). Houdaigui (2007) found a necessary number of 445 grains per RVE for determining the effective elastic properties of polycrystalline copper with an acceptable error of 1 %, where already optimal boundary conditions and the ergodicity hypothesis have been used. For plastic properties, the required RVE size is much larger. For example, consider an isotropic plastic matrix with stiff, isotropic, spherical inclusions of equal size. For the elastic properties, it is sufficient to have the RVE volume approximately 16 times the volume of a single inclusion. To get a reasonable estimate of the effective plastic properties, the RVE volume should be at least four times bigger than that (Glüge et al., 2012).

Multiscale FEM

The advantage of numerical homogenization is that arbitrarily complex microstructures and complicated micro-scale material laws RVE can be examined. From these, one gets discrete effective stresses along loading paths. For linear elasticity, the discrete results are sufficient. The overall effective elasticity is also linear, and one can assemble an effective stiffness on the macroscale from a few RVE calculations. However, in plasticity, the results cannot be condensed easily into an effective law. To use the results anyway in macroscale simulations, it has been proposed to couple macro and microscale finite element models (Feyel, 1999; Schröder, 2014).

It is clear that a straightforward multiscale FEM is numerically costly, and there is a lot of optimization potential. For example, effective results for specific loading paths can be stored in a database to avoid repeated evaluation of identical or similar RVE calculations (Klusemann and Ortiz, 2015). Another data reduction approach is to discretize the space of all deformations into modes, referred to as nonuniform transformation fields by Michel and Suquet (2003). The effective RVE-response to these deformation modes is determined once; see, e. g., Fritzen and Böhlke (2010b), Wulfinghoff and Reese (2016). At runtime of the macroscale FE model, the local effective material is approximated by the superposition of the stored values that comprise the local deformation.

Further, the RVE calculations can be made more efficient by exploiting approximate ergodicity (e. g., Huet, 1990; Hazanov and Huet, 1994), which replace a large RVE with an ensemble averaging over several smaller RVEs. This is numerically favorable because the numerical effort grows superlinear with the RVE size, but linearly with the number of RVE realizations. Other approaches are to optimize the representativity of the microstructure (e. g., Schröder et al., 2011) as well as the boundary conditions and the shape of the RVE (e. g., Fritzen and Böhlke, 2010a; Glüge et al., 2012; Glüge and Weber, 2013; Glüge, 2013).

One can see that not only material scientists, engineers and physicists but also IT and numerics experts are needed for numerical homogenization. Further information can be found in the literature, such as Llorca et al. (2007), Efendiev and Hou (2009), Zohdi and Wriggers (2008), Yvonnet (2019).

The remainder of this book is dedicated to analytical homogenization methods.

5 Homogenization with the aid of structural mechanics

Some porous or foam-like microstructures can be approximated with structural mechanics models such as beams and plates. The RVEs are then made of such structures, which may be solvable analytically. We will examine simple RVEs made of beams for a small strain-and a large strain homogenization task.

5.1 Small strains: a simple model for honeycombs

We consider a honeycomb structure; see Figure 5.1. It is modeled as an RVE made of Bernoulli[1]–Euler beams, which can be examined with standard methods from basic mechanics; see also Gibson and Ashby (1999), Section 4.3. We consider the plane problem with the following assumptions: The connection points are stiff, such that the angle between two beams remains 120° in the deformed state; see Figures 5.2 and 5.3. Further, the deformations are small, so the nominal stress is close to the true stress. With the periodicity, it is clear that the beams deform symmetrically, as sketched in

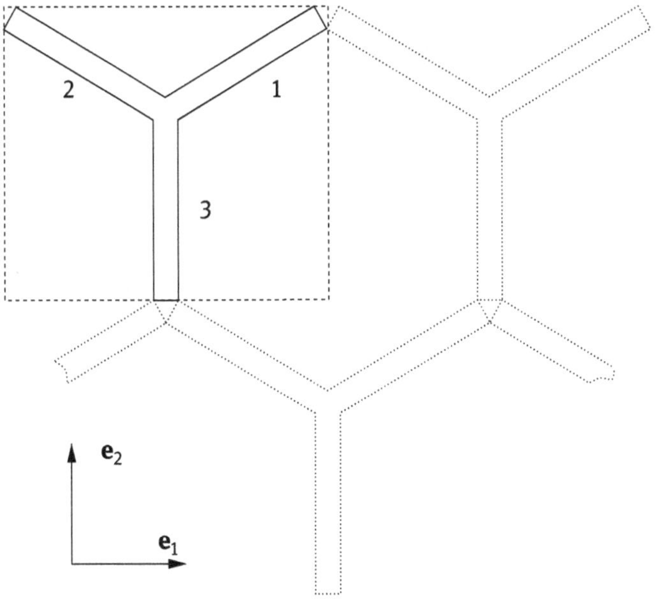

Figure 5.1: Honeycomb structure consisting of three beams arranged in a Y-shaped RVE of thickness *d* perpendicular to the paper plane.

1 Jakob I Bernoulli, 1655–1705 (https://en.wikipedia.org/wiki/Jakob_I_Bernoulli).

https://doi.org/10.1515/9783110793529-005

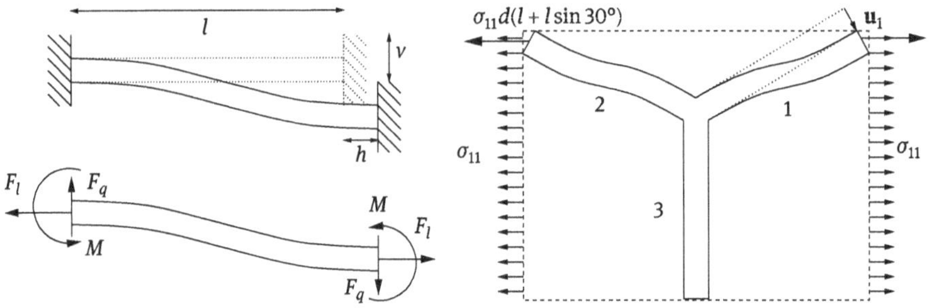

Figure 5.2: Left: Symmetrically deformed beam, clamped on both ends. The internal forces are defined such that $M > 0$, $F_q > 0$ and $F_l > 0$ when $v > 0$. Right: Honeycomb-RVE under tension in horizontal direction $\mathbf{e}_1$.

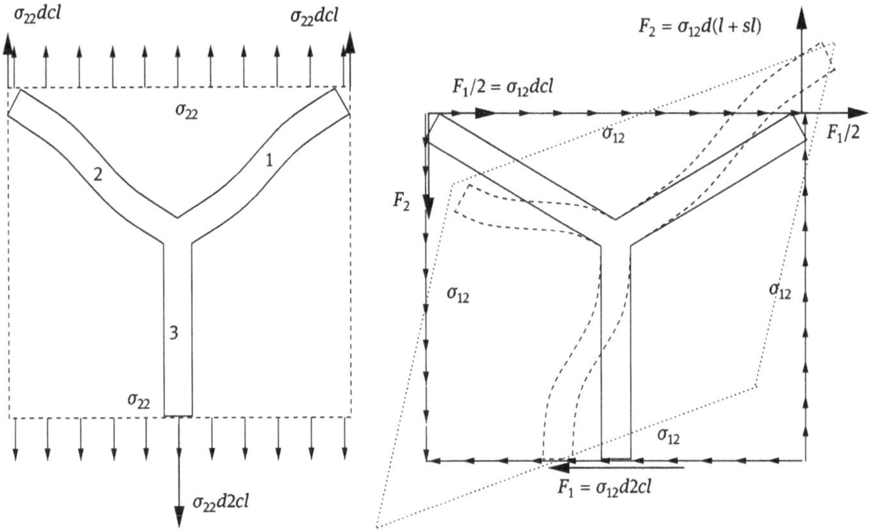

Figure 5.3: Left: RVE under tension in vertical $\mathbf{e}_2$-direction. Right: RVE under shear loading.

Figure 5.2. We use beam theory to obtain a relation between displacements and internal forces. The differential equation for the bending line along the beam coordinate s is

$$EIw''''(s) = q(s), \tag{5.1}$$

with the bending stiffness EI. With a vanishing line load $q(s) = 0$, the bending line is a cubic polynomial $w(s) = c_3s^3 + c_2s^2 + c_1s + c_0$. The shear force is

$$F_q(s) = -EIw'''(s) = -6EIc_3. \tag{5.2}$$

The parameters $c_{0...3}$ depend on the location of the origin of the coordinate s. To make use of the symmetry of the bending it is advantageous to place the coordinate system in the beam center, such that the even part of $w(s)$ is zero, hence $c_2 = 0$ and $c_0 = 0$, while c_3 and c_1 can be determined from $w(l/2) = v/2$ and $w'(l/2) = 0$. One finds

$$F_q = \frac{12EI}{l^3} v \quad \leftrightarrow \quad v = \frac{l^3}{12EI} F_q. \tag{5.3}$$

The relation between the length change h and the normal force F_l is

$$F_l = \frac{EA}{l} h \quad \leftrightarrow \quad h = \frac{l}{EA} F_l, \tag{5.4}$$

with the tensile stiffness EA. We can now determine the effective stiffness of the honeycomb structure:

1. Apply the principle of sections to the RVE (cut free) and apply the base stress states $\sigma_{11}, \sigma_{22}, \sigma_{12}$.
2. Calculate the resulting section forces (internal forces) from the stresses on the corresponding beam ends.
3. Calculate the beam end's displacements using equations (5.3) and (5.4).
4. Calculate the effective strains of the RVE.
5. Since all these are linear operations, we obtain a linear relationship between the effective stresses and the effective strains. A comparison of coefficients allows then to identify the effective compliance tetrad of the homogenized material.

A reversal of these steps, namely to prescribe effective strains and identify effective stresses, is also possible; see Section 5.2.

Uniaxial tension in horizontal direction

In Figure 5.2, a tensile loading in the horizontal direction along with the expected deformation is sketched. For decomposing forces and displacements into components parallel and perpendicular to a beam, we often need the following abbreviations:

$$c = \cos 30° = \sqrt{3}/2 \tag{5.5}$$

$$s = \sin 30° = 1/2 \quad \text{(the beam coordinate } s \text{ is no longer needed).} \tag{5.6}$$

The horizontal force is

$$F = \sigma_{11}(l + sl)d, \tag{5.7}$$

with the thickness d perpendicular to the paper plane. Decomposing F into parts parallel and perpendicular to the beam F_l and F_q, applying $v(F_q)$ and $h(F_l)$ (equations (5.3) and (5.4)) and projecting into the horizontal $\mathbf{e}_1$- and vertical $\mathbf{e}_2$-direction gives

$$\mathbf{u}_1 = (sv(sF) + ch(cF))\mathbf{e}_1 + (sh(sF) - cv(cF))\mathbf{e}_2 \tag{5.8}$$

$$\mathbf{u}_2 = -(sv(sF) + ch(cF))\mathbf{e}_1 + (sh(sF) - cv(cF))\mathbf{e}_2 \tag{5.9}$$
$$\mathbf{u}_3 = \mathbf{o}. \tag{5.10}$$

The nominal effective strain of the RVE is the relative length change over the initial length,

$$\varepsilon_{11} = \frac{2u_{1,1}}{2cl} \tag{5.11}$$
$$\varepsilon_{22} = \frac{u_{1,2}}{l + sl}. \tag{5.12}$$

Uniaxial tension in vertical direction

In Figure 5.3, a tensile loading in the vertical direction along with the expected deformation is sketched. The resulting force on the bottom is

$$F = \sigma_{22} d2cl, \tag{5.13}$$

on the top this force is distributed equally. As in the first case, we decompose F on the top side and denote the displacements of the beam ends w. r. t. the center point:

$$\mathbf{u}_1 = (-sv(cF/2) + ch(sF/2))\mathbf{e}_1 + (sh(sF/2) + cv(cF/2))\mathbf{e}_2 \tag{5.14}$$
$$\mathbf{u}_2 = -(-sv(cF/2) + ch(sF/2))\mathbf{e}_1 + (sh(sF/2) + cv(cF/2))\mathbf{e}_2 \tag{5.15}$$
$$\mathbf{u}_3 = -h(F)\mathbf{e}_2 \tag{5.16}$$

The effective nominal strain is the relative length change over the initial length,

$$\varepsilon_{11} = \frac{2u_{1,1}}{\sqrt{3}l} \tag{5.17}$$
$$\varepsilon_{22} = \frac{u_{1,2} - u_{3,2}}{3l/2}. \tag{5.18}$$

Identification of the effective elasticity

The linear relation between stresses and strains can now be summarized:

$$\begin{bmatrix} \varepsilon_{11} \\ \varepsilon_{22} \end{bmatrix} = \begin{bmatrix} \frac{3^{3/2}l}{4EA} + \frac{l^3}{16\sqrt{3}EI} & \frac{\sqrt{3}l}{4EA} - \frac{l^3}{16\sqrt{3}EI} \\ \text{sym} & \frac{3^{3/2}l}{4EA} + \frac{l^3}{16\sqrt{3}EI} \end{bmatrix} \begin{bmatrix} \sigma_{11} \\ \sigma_{22} \end{bmatrix}. \tag{5.19}$$

It may be surprising that $S_{1111} = S_{2222}$, because the tensile direction is in one case perpendicular to one of the hexagon's sides and parallel to one of the hexagon's sides in the other case. This can only be understood from a more general perspective of symmetry transformations. The structure has a 60° rotational symmetry. The elasticity tensor needs to satisfy this symmetry. Since it is a fourth-order tensor, at most a four-fold symmetry can be distinguished; see Section 13.3. Therefore, higher-fold symmetries than four collapse to isotropic elasticity, which has two independent elasticity constants. In the inverse Hookean law, Young's modulus and Poisson's[2] ratio can be identified easily,

$$E = S_{1111}^{-1} = \frac{16\sqrt{3}EAEI}{EAl^3 + 36EIl} \tag{5.20}$$

$$\nu = -\varepsilon_{22}/\varepsilon_{11} = -S_{2211}/S_{1111} = 1 - \frac{48EI}{36EI + EAl^2}. \tag{5.21}$$

It is noteworthy that Poisson's ratio ν falls into the interval $-1/3 < \nu < 1$ for the extremal cases $EAl^2 \ll EI$ and $EI \ll EAl^2$. One can now examine the effect of the beam stiffnesses EA and EI, which can be adjusted to some extent independently by varying the beam's cross-sections. If, for example, a small ν is desired, $I \ll Al^2$ is a design goal.

The effective shear modulus G can already be given in terms of E and ν,

$$G = \frac{E}{2(1+\nu)} = \frac{4\sqrt{3}EAEI}{12EIl + EAl^3}; \tag{5.22}$$

see, e. g., Bertram and Glüge (2015) (conversion table at the end of Section 4.1.2). We can complete with $S_{1212} = 1/G$ and $S_{1112} = S_{2212} = 0$ the effective elasticity law,

$$\begin{bmatrix} \varepsilon_{11} \\ \varepsilon_{22} \\ \varepsilon_{12} \end{bmatrix} = \begin{bmatrix} \frac{3^{3/2}l}{4EA} + \frac{l^3}{16\sqrt{3}EI} & \frac{\sqrt{3}l}{4EA} - \frac{l^3}{16\sqrt{3}EI} & 0 \\ & \frac{3^{3/2}l}{4EA} + \frac{l^3}{16\sqrt{3}EI} & 0 \\ \text{sym} & & \frac{12EIl+EAl^3}{4\sqrt{3}EAEI} \end{bmatrix} \begin{bmatrix} \sigma_{11} \\ \sigma_{22} \\ \sigma_{12} \end{bmatrix}. \tag{5.23}$$

We show next that analyzing a shear deformation of the RVE yields the same result.

Shear

In Figure 5.3, a shear loading along with the expected deformation is sketched. For the displacements of the endpoints, we find

$$\mathbf{u}_1 = (-s\nu(cF_2 - sF_1/2) + cu(sF_2 + cF_1/2))\mathbf{e}_1 + (c\nu(cF_2 - sF_1/2) + su(sF_2 + cF_1/2))\mathbf{e}_2 \tag{5.24}$$

$$\mathbf{u}_2 = (-s\nu(cF_2 - sF_1/2) + cu(sF_2 + cF_1/2))\mathbf{e}_1 + (-c\nu(cF_2 - sF_1/2) - su(sF_2 + cF_1/2))\mathbf{e}_2 \tag{5.25}$$

$$\mathbf{u}_3 = -\nu(F_1)\mathbf{e}_1, \tag{5.26}$$

with $F_1 = \sigma_{12}d2cl$ and $F_2 = \sigma_{12}d(l + sl)$. The strain component ε_{12} is defined as

2 Siméon Denis Poisson, 1781–1840 (https://en.wikipedia.org/wiki/Sim%C3%A9on_Denis_Poisson).

$$\varepsilon_{12} = \frac{1}{2}\left(\frac{\partial u_1}{\partial x_2} + \frac{\partial u_2}{\partial x_1}\right). \tag{5.27}$$

The finite displacement version is

$$\varepsilon_{12} = \frac{1}{2}\left(\frac{\Delta u_1}{\Delta x_2} + \frac{\Delta u_2}{\Delta x_1}\right). \tag{5.28}$$

For the first quotient of differences, we consider the ends of beams 1 and 3. For the second quotient, we consider the ends of beams 1 and 2,

$$\varepsilon_{12} = \frac{1}{2}\left(\frac{u_{1,1} - u_{3,1}}{l + sl} + \frac{u_{1,2} - u_{2,2}}{2cl}\right) = \frac{12EIl + EAl^3}{4\sqrt{3}EAEI}\sigma_{12}. \tag{5.29}$$

This corresponds to the value that we have already found by using the isotropy. The effective strains ε_{11} and ε_{22} are zero. The calculations in this section are summarized in the Mathematica notebook in Listing 5.1.

5.2 Large strains: a nonlinear truss model

The approach above does not withstand a rigorous examination and works only because of the symmetry. Otherwise, the localization of stresses to internal forces is not that simple. For example, if the beams have different cross-sections, the center point would move, and it would be necessary to solve the equilibrium equations inside the unit cell to relate the internal forces to the displacements.

Moreover, it would be more convenient to denote the material law directly in the usual way $\overline{\mathbf{P}}(\overline{\mathbf{F}})$, i. e., the stresses depending on the strains (effective first Piola–Kirchhoff stresses as a function of the deformation gradient).

A better proceeding is:

1. Imposing displacements as homogeneous strain boundary conditions $\mathbf{u}_i = \overline{\mathbf{H}}\mathbf{x}_{0i}$ or equivalently $\mathbf{x}_i = \overline{\mathbf{F}}\mathbf{x}_{0i}$ on the beam ends. This is the localization of $\overline{\mathbf{F}}$ to the local quantities $\mathbf{u}_i$. It is more complicated if nonhomogeneous boundary conditions are chosen.
2. Evaluating equilibrium conditions inside the RVE in the deformed state.
3. Evaluating the resulting internal forces $\mathbf{f}_i$ at the beam endpoints, where $\mathbf{u}_i$ is prescribed.
4. Summing the discrete variant of equation (4.21) over the boundary points to get the effective Piola–Kirchhoff stresses,

$$\overline{\mathbf{P}} = \frac{1}{V_0}\sum \mathbf{f}_i \otimes \mathbf{x}_{0i}. \tag{5.30}$$

   This is the homogenization step.

Listing 5.1: Mathematica notebook (https://gitlab.com/gluegerainer/listings-homogenization-methods/-/blob/main/Listing-03_en.nb) for homogenization of the honeycomb structure using beam theory.

```
Remove["Global`*"]
(* Create empty compliance matrix *)
SS = Table[0, 3, 3] ;
(* Abbreviations *)
v[Fq_] = Fq L^3/12/EI;
h[Fl_] = L/EA Fl;
c = Sqrt[3]/2;
s = 1/2;
(* Sigma 11 *)
F = sigma11 3 L/2;
DH = FullSimplify[
    - c v[s F] (* Vertical displacements of beams 1 and 2 projected in y direction *)
    + s h[c F] (* Horizontal displacements of beams 1 and 2 projected in y direction *)    ];
DB = FullSimplify[
    + 2 s v[s F] (* Vertical displacements of beams 1 and 2 projected in x direction *)
    + 2 c h[c F] (* Horizontal displacements of beams 1 and 2 projected in x direction *)    ];
eps11 = FullSimplify[DB/(2 c L)];
eps22 = FullSimplify[DH/(L + s L)];
SS[[1, 1]] = eps11/sigma11;
SS[[1, 2]] = eps22/sigma11;
(* Sigma 22 *)
F = sigma22 Sqrt[3] L;
DH = FullSimplify[
    + h[F]       (* Elongation of beam 3 *)
    + c v[c F/2] (* Vertical displacements of beams 1 and 2 projected in y direction *)
    + s h[s F/2] (* Horizontal displacements of beams 1 and 2 projected in y direction *)    ];
DB = FullSimplify[
    - 2 s v[c F/2] (* Vertical displacements of beams 1 and 2 projected in x direction *)
    + 2 c h[s F/2] (* Horizontal displacements of beams 1 and 2 projected in x direction *) ];
eps11 = FullSimplify[DB/(2 c L)];
eps22 = FullSimplify[DH/(L + s L)];
SS[[2, 1]] = eps11/sigma22;
SS[[2, 2]] = eps22/sigma22;
(* Sigma12 *)
F1 = sigma12 2 c L;
F2 = sigma12 (L + s L);
u1 = { - s v[c F2 - s F1/2] + c h[s F2 + c F1/2],
       + c v[c F2 - s F1/2] + s h[s F2 + c F1/2]};
u2 = { - s v[c F2 - s F1/2] + c h[s F2 + c F1/2],
       - c v[c F2 - s F1/2] - s h[s F2 + c F1/2]};
u3 = { - v[F1], 0};
eps12 = FullSimplify[(u1[[2]] - u2[[2]])/(Sqrt[3] L) + (u1[[1]] - u3[[1]])/(L + s L)];
SS[[3, 3]] = eps12/sigma12;
SS // MatrixForm
EMO = 1/SS[[1, 1]] // FullSimplify
nu = -SS[[1, 2]]/SS[[1, 1]] // FullSimplify
G = EMO/2/(1 + nu) // FullSimplify
```

This approach leads directly to an effective material law of the form $\overline{\mathbf{P}}(\overline{\mathbf{F}})$. Using Nanson's[3] formula for converting differential area elements between two configurations, we obtain the Cauchy stresses

$$\overline{\boldsymbol{\sigma}} = \overline{J}^{-1}\overline{\mathbf{P}}(\overline{\mathbf{F}})\,\overline{\mathbf{F}}^{T}, \quad \overline{J} = \det(\overline{\mathbf{F}}), \tag{5.31}$$

which becomes with $\overline{J} = V/V_0$ and $\mathbf{x}_i = \overline{\mathbf{F}}\mathbf{x}_{0i}$

$$\overline{\boldsymbol{\sigma}} = \frac{1}{V}\sum \mathbf{f}_i \otimes \mathbf{x}_i. \tag{5.32}$$

Example

We consider a microstructure made of linear springs that form equilateral triangles; see Figure 5.4. The springs have stiffness c and the force-free length l_0. In this example, we have no effort inside the RVE, since the position of the boundary points fully controls its deformation. We start by imposing the displacement with the coordinate system in the lower left corner,

$$\mathbf{x}_i = \overline{\mathbf{F}}\mathbf{x}_{0i}, \quad \text{with} \quad \mathbf{x}_{01} = \mathbf{o}, \quad \mathbf{x}_{02} = l_0\mathbf{e}_1, \quad \mathbf{x}_{03} = l_0(1/2\mathbf{e}_1 + \sqrt{3}/2\mathbf{e}_2). \tag{5.33}$$

The length change of the springs is

$$\Delta l_{12} = \|\mathbf{x}_2 - \mathbf{x}_1\| - l_0, \tag{5.34}$$

$$\Delta l_{23} = \|\mathbf{x}_3 - \mathbf{x}_2\| - l_0, \tag{5.35}$$

$$\Delta l_{31} = \|\mathbf{x}_1 - \mathbf{x}_3\| - l_0, \tag{5.36}$$

which gives the internal forces $\mathbf{f}_{ij}$ in the direction of the spring's axes,

$$\mathbf{f}_{12} = c\Delta l_{12}\frac{\mathbf{x}_2 - \mathbf{x}_1}{\|\mathbf{x}_2 - \mathbf{x}_1\|}, \tag{5.37}$$

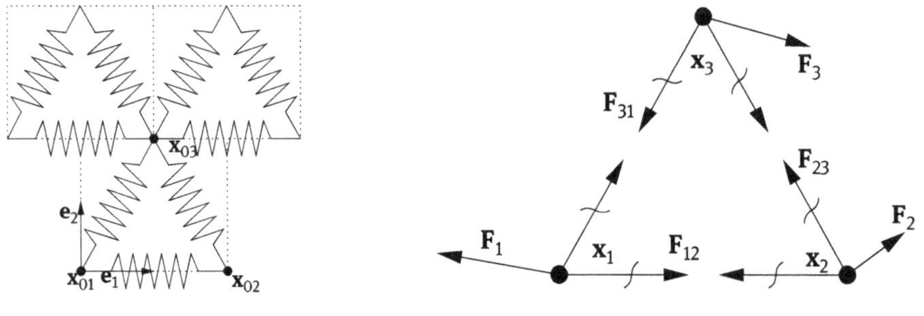

Figure 5.4: Left: The microstructure is made of springs of stiffness c and length l_0 that form equilateral triangles. Right: Internal forces $\mathbf{F}_{12}$, $\mathbf{F}_{23}$ and $\mathbf{F}_{31}$.

3 Edward J. Nanson, 1850–1936 (https://en.wikipedia.org/wiki/Edward_J._Nanson).

$$\mathbf{f}_{23} = c\Delta l_{23}\frac{\mathbf{x}_3 - \mathbf{x}_2}{\|\mathbf{x}_3 - \mathbf{x}_2\|}, \tag{5.38}$$

$$\mathbf{f}_{31} = c\Delta l_{31}\frac{\mathbf{x}_1 - \mathbf{x}_3}{\|\mathbf{x}_1 - \mathbf{x}_3\|}. \tag{5.39}$$

We choose the sides with the points i and $i+1 \bmod 3$ as positive and negative sides of the section, as shown in Figure 5.4. Therefore, the tensile forces point to the joint with the higher index. The resulting forces at the points $\mathbf{x}_1$, $\mathbf{x}_2$, $\mathbf{x}_3$ are therefore

$$\mathbf{f}_1 = \mathbf{f}_{31} - \mathbf{f}_{12}, \tag{5.40}$$

$$\mathbf{f}_2 = \mathbf{f}_{12} - \mathbf{f}_{23}, \tag{5.41}$$

$$\mathbf{f}_3 = \mathbf{f}_{23} - \mathbf{f}_{31}. \tag{5.42}$$

We can now write down the Piola–Kirchhoff stresses. The continuous averaging rule is

$$\overline{\mathbf{P}} = \frac{1}{V_0}\int_{\partial\Omega} \mathbf{t} \otimes \mathbf{x}_0 \, dA_0, \tag{5.43}$$

which becomes

$$\overline{\mathbf{P}} = \frac{1}{V_0}\sum \mathbf{t}_i \otimes \mathbf{x}_{0i}\Delta A_{0i} \tag{5.44}$$

$$= \frac{1}{V_0}\sum \frac{1}{A_i}\mathbf{f}_i \otimes \mathbf{x}_{0i}A_{0i} \tag{5.45}$$

$$= \frac{1}{V_0}\sum \frac{A_{0i}}{A_i}\mathbf{f}_i \otimes \mathbf{x}_{0i} \tag{5.46}$$

upon discretization, with the undeformed RVE volume V_0. In our 2D case, this is the area $l_0^2\sqrt{3}/2$ of the unit cell. Pulling back the right entry of $\overline{\mathbf{P}}$ with $\overline{J}^{-1}\overline{\mathbf{F}}^T$ leads with $\mathbf{x}_{0i} = \overline{\mathbf{F}}^{-1}\mathbf{x}_i$ to the Cauchy stresses,

$$\overline{\boldsymbol{\sigma}} = \overline{J}^{-1}\overline{\mathbf{P}}\,\overline{\mathbf{F}}^T = \frac{1}{V_0\overline{J}}(\mathbf{f}_1 \otimes \mathbf{x}_1\overline{\mathbf{F}}^{-T} + \mathbf{f}_2 \otimes \mathbf{x}_2\overline{\mathbf{F}}^{-T} + \mathbf{f}_3 \otimes \mathbf{x}_3\overline{\mathbf{F}}^{-T})\overline{\mathbf{F}}^T \tag{5.47}$$

$$= \frac{1}{V}(\mathbf{f}_1 \otimes \mathbf{x}_1 + \mathbf{f}_2 \otimes \mathbf{x}_2 + \mathbf{f}_3 \otimes \mathbf{x}_3). \tag{5.48}$$

Substituting the forces and the force-displacement relation of the springs give

$$\overline{\boldsymbol{\sigma}} = \frac{1}{V}((\mathbf{f}_{31} - \mathbf{f}_{12}) \otimes \mathbf{x}_1 + (\mathbf{f}_{12} - \mathbf{f}_{23}) \otimes \mathbf{x}_2 + (\mathbf{f}_{23} - \mathbf{f}_{31}) \otimes \mathbf{x}_3) \tag{5.49}$$

$$= \frac{1}{V}(\mathbf{f}_{31} \otimes (\mathbf{x}_1 - \mathbf{x}_3) + \mathbf{f}_{12} \otimes (\mathbf{x}_2 - \mathbf{x}_1) + \mathbf{f}_{23} \otimes (\mathbf{x}_3 - \mathbf{x}_2)) \tag{5.50}$$

$$= \frac{1}{V}\Bigg(\frac{c\Delta l_{31}}{\|\mathbf{x}_1 - \mathbf{x}_3\|}(\mathbf{x}_1 - \mathbf{x}_3) \otimes (\mathbf{x}_1 - \mathbf{x}_3)$$

$$+ \frac{c\Delta l_{12}}{\|\mathbf{x}_2 - \mathbf{x}_1\|}(\mathbf{x}_2 - \mathbf{x}_1) \otimes (\mathbf{x}_2 - \mathbf{x}_1)$$

$$+\frac{c\Delta l_{23}}{\|\mathbf{x}_3-\mathbf{x}_2\|}(\mathbf{x}_3-\mathbf{x}_2)\otimes(\mathbf{x}_3-\mathbf{x}_2)\Bigg), \tag{5.51}$$

which can be written with $\Delta\mathbf{x}_{ji}=\mathbf{x}_i-\mathbf{x}_j$ and $l_{ij}=\|\Delta\mathbf{x}_{ij}\|$ a little bit more compact,

$$\overline{\boldsymbol{\sigma}}=\frac{c}{V}\left(\frac{l_{31}-l_0}{l_{31}}\Delta\mathbf{x}_{31}\otimes\Delta\mathbf{x}_{31}+\frac{l_{12}-l_0}{l_{12}}\Delta\mathbf{x}_{12}\otimes\Delta\mathbf{x}_{12}+\frac{l_{23}-l_0}{l_{23}}\Delta\mathbf{x}_{23}\otimes\Delta\mathbf{x}_{23}\right). \tag{5.52}$$

The fractions $(l_{ij}-l_0)/l_{ij}$ correspond to the length change Δl_{ij} over the current length l_{ij}. One can see that $\overline{\boldsymbol{\sigma}}$ is symmetric. We can replace $\mathbf{x}_i$ again by $\overline{\mathbf{F}}\mathbf{x}_{0i}$,

$$\overline{\boldsymbol{\sigma}}=\frac{c}{V}\overline{\mathbf{F}}\left(\frac{l_{31}-l_0}{l_{31}}\Delta\mathbf{x}_{0,31}\otimes\Delta\mathbf{x}_{0,31}+\frac{l_{12}-l_0}{l_{12}}\Delta\mathbf{x}_{0,12}\otimes\Delta\mathbf{x}_{0,12}+\frac{l_{23}-l_0}{l_{23}}\Delta\mathbf{x}_{0,23}\otimes\Delta\mathbf{x}_{0,23}\right)\overline{\mathbf{F}}^T. \tag{5.53}$$

It is apparent that $\overline{\boldsymbol{\sigma}}(\overline{\mathbf{F}})$ is invariant under superimposed rotations,

$$\overline{\boldsymbol{\sigma}}(\mathbf{Q}\overline{\mathbf{F}})=\mathbf{Q}\overline{\boldsymbol{\sigma}}(\overline{\mathbf{F}})\mathbf{Q}^T \quad \forall\mathbf{Q}\in\mathrm{SO}(3); \tag{5.54}$$

see, e. g., Glüge (2018), since the length l_{ij} are invariant under rotations $\mathbf{Q}$. The obtained expression is a nonlinear, anisotropic tensor function with anisotropic structural tensors $\Delta\mathbf{x}_{0,ij}\otimes\Delta\mathbf{x}_{0,ij}$, in contrast to the isotropic effective elasticity of the honeycomb. This is a nice example of how a nonlinear, large strain modeling captures an anisotropy that is not accounted for in a linear, small strain setting. The effective nonlinear stress-strain relation contains tensors of arbitrary even order, such that all n-fold symmetries n are included; see Section 13.3. The anisotropy is somewhat weak. Comparing the deformations $\overline{\mathbf{F}}=2\mathbf{e}_1\otimes\mathbf{e}_1+\mathbf{e}_2\otimes\mathbf{e}_2$ and $\overline{\mathbf{F}}=\mathbf{e}_1\otimes\mathbf{e}_1+2\mathbf{e}_2\otimes\mathbf{e}_2$, i. e., 100 % nominal strain without lateral contraction in horizontal and vertical direction, one finds Cauchy stresses that differ by approximately 7.4 %.

A Mathematica notebook that performs these calculations and displays the results graphically is given in Listing 5.2 and Figure 5.5 along with its output. In Listing 5.3, the resulting nonlinear law is linearized by inserting $\mathbf{F}=\mathbf{I}+\boldsymbol{\varepsilon}$ and linearizing $\boldsymbol{\sigma}$ at $\boldsymbol{\varepsilon}=0$ in $\boldsymbol{\varepsilon}$. One obtains as the stiffness matrix

$$C_{ij}=\begin{bmatrix}3^{3/2}c/4 & \sqrt{3}c/4 & 0\\ & 3^{3/2}c/4 & 0\\ \text{sym} & & \sqrt{3}c\end{bmatrix}\mathbf{E}_i\otimes\mathbf{E}_j,\quad i,j\in\{1,2,4\}; \tag{5.55}$$

see equation (2.12) in Section 2.3 for the definition of the basis $\mathbf{E}_i$. One can see that $C_{44}=2(C_{11}-C_{12})$, i. e., that the result is isotropic.

Moreover, it is noteworthy that the length scale l_0 does not appear in the resulting effective elasticity laws, and that the stresses have the physical unit of a line load (or spring constant) N/m, since we consider 2D problems. Both holds also for the nonlinear version. One can easily check that the final expression for $\overline{\boldsymbol{\sigma}}$ (equations (5.52) or (5.55)) is scale invariant w. r. t. l_0.

Listing 5.2: Mathematica notebook (https://gitlab.com/gluegerainer/listings-homogenization-methods/-/blob/main/Listing-04_en.nb) for nonlinear homogenization of the spring model in Figure 5.4.

```
Remove["Global`*"]
c = 2; (* spring constant in N/m *)
l0 = 1; (* edge length in m *)
x0 = {{0, 0}, {l0, 0}, { l0/2, Sqrt[3] l0/2}}; (* coordinates of the end points, undeformed *)
V0 = l0^2 Sqrt[3]/2;  (* RVE volume = area of the unit cell *)
Manipulate[
 F = {{F11, F12}, {F21, F22}}; (* effective deformation gradient *)
 x = Table[F.x0[[i]], {i, 1, 3}]; (* coordinates of the end points, deformed *)
 Forces = {c (Norm[x[[1]] - x[[3]]] - l0) Normalize[x[[1]] - x[[3]]], (* spring forces *)
           c (Norm[x[[2]] - x[[1]]] - l0) Normalize[x[[2]] - x[[1]]],
           c (Norm[x[[3]] - x[[2]]] - l0) Normalize[x[[3]] - x[[2]]]};
 RForces = {Forces[[1]] - Forces[[2]], Forces[[2]] - Forces[[3]], Forces[[3]] - Forces[[1]]}; (*
       internal forces *)
 P = 1/V0 Sum[ Outer[Times, RForces[[i]], x0[[i]]], {i, 1, 3}]; (* Piola-Kirchhoff-stresses *)
 sigma = P.Transpose[F]/Det[F]; (* Cauchy-stresses *)
 es = Eigensystem[sigma]; (* Eigenvalues and eigenvectors of the Cauchy-stresses *)
 center = (x[[1]] + x[[2]] + x[[3]])/3; (* graphical output *)
 {Graphics[{
    Line[{x0[[1]], x0[[2]], x0[[3]], x0[[1]]}], (* undeformed placement*)
    Red, Line[{x[[1]], x[[2]], x[[3]], x[[1]]}], (* deformed placement *)
    Arrow[{ (* internal forces *)
      {x[[1]], x[[1]] + RForces[[1]]},
      {x[[2]], x[[2]] + RForces[[2]]},
      {x[[3]], x[[3]] + RForces[[3]]}}],
    Arrowheads[{-.05, .05}], (* spectral representation of the Cauchy stresses *)
    Green,
    Arrow[{center - Normalize[es[[2, 1]]] es[[1, 1]], center + Normalize[es[[2, 1]]] es[[1,
      1]]}],
    Arrow[{center - Normalize[es[[2, 2]]] es[[1, 2]], center + Normalize[es[[2, 2]]] es[[1,
      2]]}]
    }], MatrixForm[sigma] // N}
 , {{F11, 1}, 0.5, 2}, {{F12, 0}, -2, 2}, {{F21, 0}, -2, 2}, {{F22, 1}, 0.5, 2}]
```

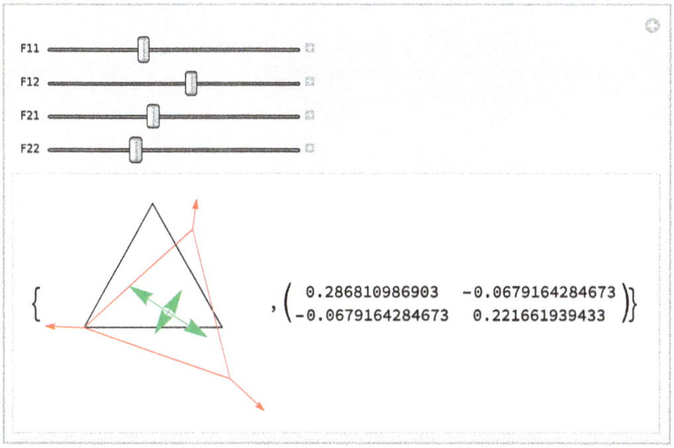

Figure 5.5: Manipulate environment in Listing 5.2 for nonlinear homogenization of the spring model in Figure 5.4.

Listing 5.3: Mathematica notebook (https://gitlab.com/gluegerainer/listings-homogenization-methods/-/blob/main/Listing-05_en.nb) for linearizing the nonlinear effective stress-strain relation of the spring structure.

```
Remove["Global`*"]
$Assumptions = {Element[{eps11, eps12, eps22, l0, c}, Reals], l0 > 0, c > 0};
x0 = {{0, 0}, {l0, 0}, { l0/2, Sqrt[3] l0/2}}; (* coordinates of the end points, undeformed *)
V0 = l0 l0 Sqrt[3]/ 2;  (* RVE volume = area of the unit cell *)
F = {{1 + eps11, eps12}, {eps12, 1 + eps22}}; (* effective deformation gradient *)
x = Table[F.x0[[i]], {i, 1, 3}]; (* coordinates of the end points, deformed *)
Forces = {c (Norm[x[[1]] - x[[3]]] - l0) Normalize[x[[1]] - x[[3]]], (* spring forces *)
          c (Norm[x[[2]] - x[[1]]] - l0) Normalize[x[[2]] - x[[1]]],
          c (Norm[x[[3]] - x[[2]]] - l0) Normalize[x[[3]] - x[[2]]]};
RForces = {Forces[[1]] - Forces[[2]], Forces[[2]] - Forces[[3]], Forces[[3]] - Forces[[1]]}; (*
      internal forces *)
P = 1/V0 Sum[Outer[Times, RForces[[i]], x0[[i]]], {i, 1, 3}]; (* Piola-Kirchhoff-stresses *)
sigma = P.Transpose[F]/Det[F];
epszero = {eps11 -> 0, eps12 -> 0, eps22 -> 0}
C1111 = D[sigma[[1, 1]], eps11] /. epszero // FullSimplify
C1122 = D[sigma[[1, 1]], eps22] /. epszero // FullSimplify
C1112 = D[sigma[[1, 1]], eps12] /. epszero // FullSimplify
C2222 = D[sigma[[2, 2]], eps22] /. epszero // FullSimplify
C2212 = D[sigma[[2, 2]], eps12] /. epszero // FullSimplify
C1212 = 2 D[sigma[[1, 2]], eps12] /. epszero // FullSimplify
```

6 Estimates in terms of volume fractions

For linear differential equations, the effective properties are defined directly over volume averages. Let $\boldsymbol{\varepsilon}(\mathbf{x})$ be a strain field of a material sample, and Hooke's law $\boldsymbol{\sigma}(\mathbf{x}_0) = \mathbb{C}(\mathbf{x}_0) : \boldsymbol{\varepsilon}(\mathbf{x}_0)$ holds. Then the effective stiffness $\mathbb{C}^*$ is defined implicitly as

$$\overline{\boldsymbol{\sigma}} = \mathbb{C}^* : \overline{\boldsymbol{\varepsilon}}. \tag{6.1}$$

The strain field is not arbitrary but the gradient of a displacement field. The stresses satisfy the balance of momentum without body forces; see Section 4.1.

6.1 Voigt–Reuss bounds

Instead of solving a boundary value problem for the displacement field, we simplify the problem by approximations to obtain analytical estimates.

Voigt mean

We assume a homogeneous deformation of the entire sample, i. e., that $\boldsymbol{\varepsilon}(\mathbf{x}_0) = \overline{\boldsymbol{\varepsilon}}$ (Voigt, 1889, 1928). Then the effective stresses are

$$\overline{\boldsymbol{\sigma}} = \frac{1}{V} \int_{\Omega} \mathbb{C}(\mathbf{x}_0) : \overline{\boldsymbol{\varepsilon}} \, \mathrm{d}v_0 \tag{6.2}$$

$$= \frac{1}{V} \int_{\Omega} \mathbb{C}(\mathbf{x}_0) \, \mathrm{d}v_0 : \overline{\boldsymbol{\varepsilon}} \tag{6.3}$$

$$= \overline{\mathbb{C}} : \overline{\boldsymbol{\varepsilon}}. \tag{6.4}$$

A comparison of coefficients with the implicit definition of $\mathbb{C}^*$ in equation (6.1) allows to identify $\mathbb{C}^*$,

$$\mathbb{C}^* \approx \overline{\mathbb{C}} =: \mathbb{C}_{\text{Voigt}}. \tag{6.5}$$

This deformation is kinematically admissible but violates the equilibrium conditions if the material is not homogeneous or some other special case applies.

Reuss mean

We assume that the sample is loaded homogeneously, i. e., that $\boldsymbol{\sigma}(\mathbf{x}_0) = \overline{\boldsymbol{\sigma}}$ (Reuss, 1929,[1] Section 5). Then we have

---

1 András Reuss, 1900–1968 (https://de.wikipedia.org/wiki/Andr%C3%A1s_Reuss_(Ingenieur)).

https://doi.org/10.1515/9783110793529-006

$$\overline{\boldsymbol{\varepsilon}} = \frac{1}{V}\int\limits_{\Omega} \mathbb{C}^{-1}(\mathbf{x}_0) : \overline{\boldsymbol{\sigma}}\, \mathrm{d}v_0 \tag{6.6}$$

$$= \frac{1}{V}\int\limits_{\Omega} \mathbb{C}^{-1}(\mathbf{x}_0)\, \mathrm{d}v_0 : \overline{\boldsymbol{\sigma}} \tag{6.7}$$

$$= \overline{\mathbb{C}^{-1}} : \overline{\boldsymbol{\sigma}}, \tag{6.8}$$

what is compared to the implicit definition of $\mathbb{C}^*$ in equation (6.1),

$$\mathbb{C}^* \approx \langle \mathbb{C}^{-1} \rangle^{-1} =: \mathbb{C}_{\text{Reuss}}. \tag{6.9}$$

The homogeneous stress field is trivially in equilibrium, but the corresponding strain field is kinematically incompatible if the material is not homogeneous (which it probably is not since homogenization methods are needed) or some other particular case applies.

6.1.1 Interpretation of the Voigt–Reuss averages

In uniaxial elasticity, the Voigt mean corresponds to a parallel connection of two springs, and the Reuss mean corresponds to a serial connection of two springs, as sketched in Figure 6.1. One finds the corresponding rules in the theory of laminates made of isotropic materials. Depending on the shear direction, the effective shear modulus is the arithmetic mean (Voigt) of the phase's shear moduli when the shear plane normal and the shear direction lie in the laminate plane, or the harmonic mean of the phase's shear moduli (Reuss) when shearing parallel to the laminate. This is derived in detail in Section 7.1. Due to the lateral contraction, this does not hold for Young's modulus.

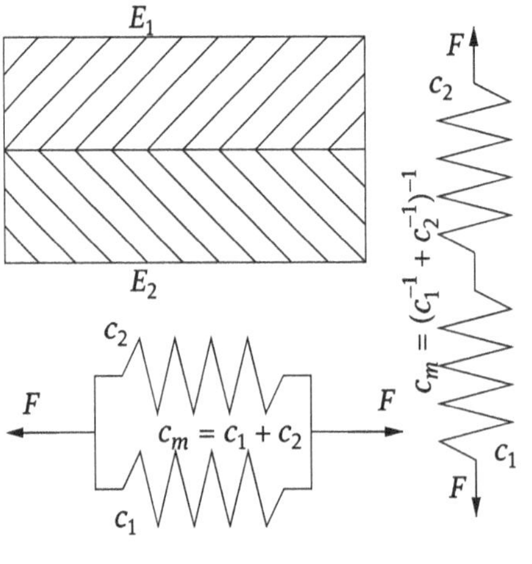

Figure 6.1: The parallel and serial connection of springs result in rules of a mixture similar to the Voigt and Reuss means.

6.1.2 Bounding properties of the Voigt and Reuss means

Voigt bound

The principle of the minimum of the elastic potential states that the potential

$$\Pi(\boldsymbol{\varepsilon}(\mathbf{x}_0)) = \int_{\Omega} \frac{1}{2}\boldsymbol{\varepsilon} : \mathbb{C} : \boldsymbol{\varepsilon} - \mathbf{b} \cdot \mathbf{u} \, dv_0 - \int_{\partial\Omega_{\text{dyn}}} \mathbf{t} \cdot \mathbf{u} \, dv_0 \tag{6.10}$$

takes a minimum at the state of equilibrium over all kinematically admissible strain fields $\boldsymbol{\varepsilon}$. All involved quantities are fields that depend on $\mathbf{x}_0$. The second integral is only evaluated over the boundary $\partial\Omega_{\text{dyn}}$ where stress boundary conditions $\mathbf{t}$ are applied.

For homogenization tasks, the body force $\mathbf{b}$ needs to be dropped. We assume linear displacement boundary conditions on the entire boundary $\mathbf{u} = \overline{\boldsymbol{\varepsilon}} \cdot \mathbf{x}_0$, which is why $\partial\Omega_{\text{dyn}} = \varnothing$, and which is why the surface integral vanishes as well. Therefore, we have

$$\int_{\Omega} \frac{1}{2} \underbrace{\boldsymbol{\varepsilon}_{\text{sol}}(\mathbf{x}_0) : \mathbb{C}(\mathbf{x}_0)}_{\boldsymbol{\sigma}_{\text{sol}}(\mathbf{x}_0)} : \boldsymbol{\varepsilon}_{\text{sol}}(\mathbf{x}_0) \, dv_0 \leq \int_{\Omega} \frac{1}{2}\boldsymbol{\varepsilon}(\mathbf{x}_0) : \mathbb{C}(\mathbf{x}_0) : \boldsymbol{\varepsilon}(\mathbf{x}_0) \, dv_0 \tag{6.11}$$

for the solution of the boundary value problem $\boldsymbol{\varepsilon}_{\text{sol}}(\mathbf{x}_0)$ and every other kinematically admissible strain field $\boldsymbol{\varepsilon}(\mathbf{x}_0)$. For the latter, we insert $\overline{\boldsymbol{\varepsilon}}$ and expand with $2/V$,

$$\frac{1}{V}\int_{\Omega} \boldsymbol{\sigma}_{\text{sol}}(\mathbf{x}_0) : \boldsymbol{\varepsilon}_{\text{sol}}(\mathbf{x}_0) \, dv_0 \leq \frac{1}{V}\int_{\Omega} \overline{\boldsymbol{\varepsilon}} : \mathbb{C}(\mathbf{x}_0) : \overline{\boldsymbol{\varepsilon}} \, dv_0. \tag{6.12}$$

With the Hill–Mandel condition (Section 4.3.3), which holds for homogeneous strain boundary conditions, we can rewrite the left-hand side. On the right-hand side, we can pull $\overline{\boldsymbol{\varepsilon}}$ out of the integral

$$\overline{\boldsymbol{\sigma}} : \overline{\boldsymbol{\varepsilon}} \leq \overline{\boldsymbol{\varepsilon}} : \frac{1}{V}\int_{\Omega} \mathbb{C}(\mathbf{x}_0) \, dv_0 : \overline{\boldsymbol{\varepsilon}}. \tag{6.13}$$

On the left side, we can insert $\overline{\boldsymbol{\sigma}} = \mathbb{C}^* : \overline{\boldsymbol{\varepsilon}}$, on the right side, we recognize the Voigt mean of the stiffness tensor (equation (6.5)),

$$\overline{\boldsymbol{\varepsilon}} : \mathbb{C}^* : \overline{\boldsymbol{\varepsilon}} \leq \overline{\boldsymbol{\varepsilon}} : \mathbb{C}_{\text{Voigt}} : \overline{\boldsymbol{\varepsilon}}, \tag{6.14}$$

or as a difference,

$$0 \leq \overline{\boldsymbol{\varepsilon}} : (\mathbb{C}_{\text{Voigt}} - \mathbb{C}^*) : \overline{\boldsymbol{\varepsilon}}. \tag{6.15}$$

Since $\overline{\boldsymbol{\varepsilon}}$ can take arbitrary values, $\mathbb{C}_{\text{Voigt}} - \mathbb{C}^*$ must be positive definite. The arithmetic mean of the local stiffness tensor $\overline{\mathbb{C}}$ is in the sense of definiteness an upper bound for the true effective stiffness $\mathbb{C}^*$.

Reuss bound

The principle of the minimum of the complementary potential states that the potential

$$\Pi^{\dagger}(\boldsymbol{\sigma}(\mathbf{x}_0)) = \int_{\Omega} \frac{1}{2}\boldsymbol{\sigma} : \mathbb{C}^{-1}\boldsymbol{\sigma}\, \mathrm{d}v_0 - \int_{\partial\Omega_{\text{kin}}} \mathbf{t}\cdot\mathbf{u}\, \mathrm{d}v_0 \tag{6.16}$$

takes a minimum at the state of equilibrium among all stress fields that satisfy $\boldsymbol{\sigma}\cdot\nabla + \rho\mathbf{b} = \mathbf{o}$. All involved quantities are fields that depend on $\mathbf{x}_0$. The second integral is evaluated only over the boundary $\partial\Omega_{\text{kin}}$ where displacement boundary conditions $\mathbf{u}$ are imposed.

With $\mathbf{b} = \mathbf{o}$ and only stress boundary conditions $\mathbf{t} = \overline{\boldsymbol{\sigma}}\cdot\mathbf{n}_0$, we can write

$$\int_{\Omega} \frac{1}{2} \boldsymbol{\sigma}_{\text{sol}}(\mathbf{x}_0) : \underbrace{\mathbb{C}^{-1}(\mathbf{x}_0) : \boldsymbol{\sigma}_{\text{sol}}(\mathbf{x}_0)}_{\boldsymbol{\varepsilon}_{\text{sol}}(\mathbf{x}_0)}\, \mathrm{d}v_0 \leq \int_{\Omega} \frac{1}{2}\boldsymbol{\sigma}(\mathbf{x}_0) : \mathbb{S}(\mathbf{x}_0) : \boldsymbol{\sigma}(\mathbf{x}_0)\, \mathrm{d}\mathbf{x}_0. \tag{6.17}$$

We insert a homogeneous stress field $\boldsymbol{\sigma}(\mathbf{x}_0) = \overline{\boldsymbol{\sigma}}$ and expand with $2/V$,

$$\frac{1}{V}\int_{\Omega} \boldsymbol{\sigma}_{\text{sol}}(\mathbf{x}_0) : \boldsymbol{\varepsilon}_{\text{sol}}(\mathbf{x}_0)\, \mathrm{d}v_0 \leq \frac{1}{V}\int_{\Omega} \overline{\boldsymbol{\sigma}} : \mathbb{C}^{-1}(\mathbf{x}_0) : \overline{\boldsymbol{\sigma}}\, \mathrm{d}v_0. \tag{6.18}$$

With the Hill–Mandel condition (Section 4.3.3), which holds for homogeneous stress boundary conditions, we can rewrite the left-hand side. On the right-hand side, we can pull $\overline{\boldsymbol{\sigma}}$ out of the integral,

$$\overline{\boldsymbol{\sigma}} : \overline{\boldsymbol{\varepsilon}} \leq \overline{\boldsymbol{\sigma}} : \frac{1}{V}\int_{\Omega} \mathbb{C}^{-1}(\mathbf{x}_0)\, \mathrm{d}v_0 : \overline{\boldsymbol{\sigma}}. \tag{6.19}$$

On the left side, we can insert $\overline{\boldsymbol{\varepsilon}} = \mathbb{C}^{*-1} : \overline{\boldsymbol{\sigma}}$, on the right side, we recognize the Reuss stiffness (equation (6.9)),

$$\overline{\boldsymbol{\sigma}} : \mathbb{C}^{*-1} : \overline{\boldsymbol{\sigma}} \leq \overline{\boldsymbol{\sigma}} : \mathbb{C}_{\text{Reuss}}^{-1} : \overline{\boldsymbol{\sigma}}, \tag{6.20}$$

or as a difference,

$$0 \leq \overline{\boldsymbol{\sigma}} : (\mathbb{C}_{\text{Reuss}}^{-1} - \mathbb{C}^{*-1}) : \overline{\boldsymbol{\sigma}}. \tag{6.21}$$

Since $\overline{\boldsymbol{\sigma}}$ can take arbitrary values, the bracket is positive definite. It is clear that an inversion of each tensor inside the bracket inverts the relation sign,

$$0 \geq \overline{\boldsymbol{\sigma}} : (\mathbb{C}_{\text{Reuss}} - \mathbb{C}^{*}) : \overline{\boldsymbol{\sigma}}. \tag{6.22}$$

Hence, we have the ordering

$$\mathbb{C}_{\text{Reuss}} \leq \mathbb{C}^{*} \leq \mathbb{C}_{\text{Voigt}}, \tag{6.23}$$

where the relation sign is to be read in the sense of definiteness as in equations (6.15) and (6.22). Therefore, the true effective elasticity lies in the interval that is spanned by the Voigt and Reuss means. Unfortunately, this interval can be large when the phase contrast is large, hence the Voigt and Reuss estimates can contain a large error; see, e. g., Figure 6.3.

It would be nice if the above inequality that holds in the sense of definiteness could be applied to the components of the stiffnesses and the other way around. Unfortunately, this is not possible. From the definiteness inequality, one can only conclude

$$C_{\text{Reuss}ii} < C^*_{ii} < C_{\text{Voigt}ii}, \quad i = 1 \dots 6 \text{ (no sum)} \tag{6.24}$$

for the components along the main diagonal w. r. t. the basis $\mathbf{E}_i \otimes \mathbf{E}_j$.

6.2 Averages between the Voigt–Reuss bounds

We next generalize to averages between the Voigt–Reuss bounds by considering

$$\mathbb{C}^*_m = \Big(\sum_{i=1\dots n} v_i \mathbb{C}_i^m\Big)^{\frac{1}{m}} \tag{6.25}$$

with the volume fractions $\sum v_i = 1$ for n phases. To maintain the physical unit of $\mathbb{C}$, the function outside the sum needs to be inverted. More generally, one would write

$$\mathbb{C}^* = g^{-1}\Big(\sum_{i=1\dots n} v_i g(\mathbb{C}_i)\Big). \tag{6.26}$$

We obtain

- The function $\mathbb{C}^*_\infty = \max(\mathbb{C}_1, \mathbb{C}_2 \dots \mathbb{C}_n)$ for $m \to \infty$ in the sense of the eigenvalues of $\mathbb{C}_i$.
- The function $\mathbb{C}^*_{-\infty} = \min(\mathbb{C}_1, \mathbb{C}_2 \dots \mathbb{C}_n)$ for $m \to -\infty$ in the sense of the eigenvalues of $\mathbb{C}_i$.
- The arithmetic mean for $m = 1$.
- The harmonic mean for $m = -1$.

In the cases $m \to \pm\infty$, the volume fractions do not enter the estimate. These are the bounds that are obtained when the volume fractions are unknown. In Figure 6.2, this asymptotic behavior is depicted along with the characteristic values $m = -1$, $m = 0$ and $m = 1$. If the volume fractions are known, it is reasonable to choose $-1 \leq m \leq 1$, since we have seen that $m = \pm 1$ gives bounds for the real stiffness. One can see that the arithmetic mean of the stiffness equals the harmonic mean of the compliance, and the other way around. The case $m = 0$ appears interesting as the canonical choice between the bounds. To examine it, we consider the limit

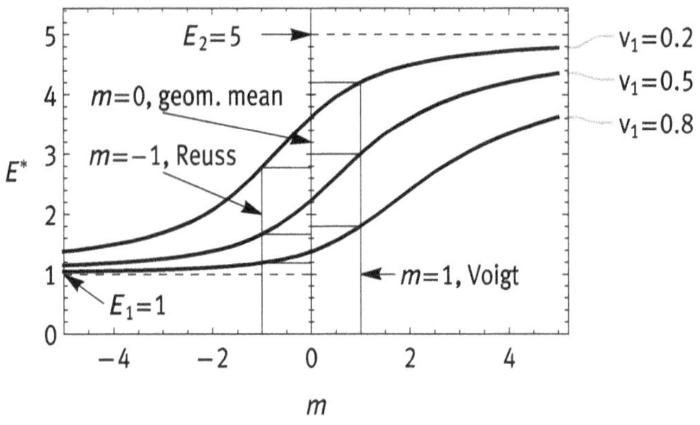

Figure 6.2: Generalized mean E^* of two moduli $E_1 = 1, E_2 = 5$ over the exponent m. For $m \to \pm\infty$, one obtains $\max(E_1, E_2)$ and $\min(E_1, E_2)$ (dashed lines), independent of volume fractions. For $m = \pm 1$, one obtains the Voigt and Reuss means. For $m = 0$, one obtains the geometric mean.

$$\lim_{m\to 0} (\ln \mathbb{C}^*_m) = \lim_{m\to 0}\left(\frac{1}{m} \ln \sum_{i=1\ldots n} v_i \mathbb{C}_i^m\right) = \lim_{m\to 0}(m^{-1} \ln \mathbf{I}), \tag{6.27}$$

which leads to the undefined quotient $\mathbf{0}/0$. The logarithm, as well as the power m, are best evaluated in the spectral form. Then we can restrict attention to scalar functions applied to the eigenvalues C_i. With Bernoulli's[2] rule for limits, we obtain the limit by individually taking the derivative w. r. t. m of the nominator and the denominator, which gives

$$\lim_{m\to 0} (\ln C^*_m) = \lim_{m\to 0} \frac{\sum_{i=1\ldots n} v_i \mathbb{C}_i^m \ln C_i}{\sum_{i=1\ldots n} v_i \mathbb{C}_i^m}. \tag{6.28}$$

For $m = 0$, this becomes

$$\ln C^*_0 = \sum_{i=1\ldots n} v_i \ln C_i. \tag{6.29}$$

Thus we obtain the geometric mean (Aleksandrov and Aisenberg, 1966; Matthies and Humbert, 1995),

$$\mathbb{C}^*_0 = \exp\left(\sum_{i=1\ldots n} v_i \ln \mathbb{C}_i\right), \tag{6.30}$$

with the mutually inverse functions ln and exp. It has the nice property of giving the result independent of whether it is applied to the stiffness $\mathbb{C}$ or the compliance $\mathbb{S} = \mathbb{C}^{-1}$.

2 Johann I Bernoulli, 1667–1748 (https://en.wikipedia.org/wiki/Johann_I_Bernoulli).

$$\mathbb{S}_0^* = \exp \sum v_i \ln \mathbb{S}_i = \exp \sum v_i \ln \mathbb{C}_i^{-1} = \exp\Big(- \sum v_i \ln \mathbb{C}_i\Big) = \Big(\exp \sum v_i \ln \mathbb{C}_i\Big)^{-1} = \mathbb{C}_0^{*-1} \tag{6.31}$$

We can summarize

- The arithmetic mean of the stiffness equals the inverse of the harmonic mean of the compliance. It is called the Voigt mean and poses an upper bound to the true effective stiffness.
- The arithmetic mean of the compliance equals the inverse of the harmonic mean of the stiffness. It is called the Reuss mean and poses an upper bound for the compliance, hence a lower bound for the stiffness.
- The geometric mean of the stiffness equals the inverse of the geometric mean of the compliance.

Hill mean

The geometric mean is not the only mean that gives the same results whether applied to the stiffness or the compliance. The same holds for the Hill mean, which is obtained by iterating between the Voigt and Reuss means:

$$\mathbb{C}_{\mathrm{Hill}k} = \left(\mathbb{C}_{\mathrm{Hill}k-1} + \mathbb{S}_{\mathrm{Hill}k-1}^{-1}\right)/2 \tag{6.32}$$

$$\mathbb{S}_{\mathrm{Hill}k} = \left(\mathbb{C}_{\mathrm{Hill}k-1}^{-1} + \mathbb{S}_{\mathrm{Hill}k-1}\right)/2. \tag{6.33}$$

It is $\mathbb{C}_{\mathrm{Hill}k} \neq \mathbb{S}_{\mathrm{Hill}k}^{-1}$. The starting values for this iteration are the Voigt and Reuss means,

$$\mathbb{C}_{\mathrm{Hill}0} = \sum v_i \mathbb{C}_i \tag{6.34}$$

$$\mathbb{S}_{\mathrm{Hill}0} = \sum v_i \mathbb{S}_i. \tag{6.35}$$

Since the kth Hill means $\mathbb{C}_{\mathrm{Hill}k}$ and $\mathbb{S}_{\mathrm{Hill}k}$ always lie between the $k-1$-Hill means, they converge against a common value

$$\mathbb{C}_{\mathrm{Hill}\infty} = \mathbb{S}_{\mathrm{Hill}\infty}^{-1}. \tag{6.36}$$

Interestingly, the converged Hill mean and the geometric mean can still differ significantly, although both have the same invariance property mentioned above. In Figure 6.3, the means are plotted for a phase contrast $G_2/G_1 = 20$. One can see nicely that

- the interval spanned by the Voigt–Reuss bounds is big and amounts to approximately half the phase contrast at $v_1 \approx 0.2$,
- the Hill mean (dark red) converges quickly to a value that is far off the geometric mean (dark green).

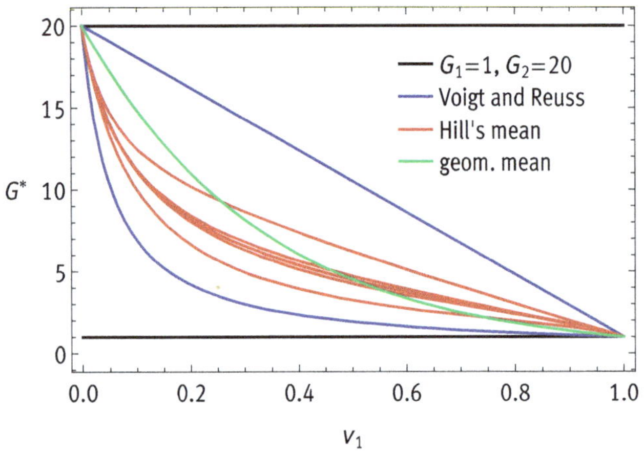

Figure 6.3: Different means in one chart. The generating Mathematica notebook is given in Listing 6.1.

Listing 6.1: Mathematica notebook (https://gitlab.com/gluegerainer/listings-homogenization-methods/-/blob/main/Listing-06_en.nb) for simple averages of dimensionless moduli 1 and 20 (see Figure 6.3).

```
Remove["Global`*"];
G1 = 1;
G2 = 20;
GV = v1 G1 + (1 - v1) G2;  (* VOIGT *)
GR = (v1 G1^(-1) + (1 - v1) G2^(-1))^(-1);  (* REUSS *)
GG = Exp[v1 Log[G1] + (1 - v1) Log[G2]];  (* GEOM *)
tbl = RecurrenceTable[{   (* Hill-Iterations *)
   GABOVE[n + 1] == (GABOVE[n] + GBELOW[n])/2,
   GBELOW[n + 1] == 2/(1/GABOVE[n] + 1/GBELOW[n]), GABOVE[1] == GV,
   GBELOW [1] == GR}, {GABOVE, GBELOW}, {n, 1, 5}];
Plot[ {G1, G2, GV, GR, tbl[[2, 1]], tbl[[2, 2]], tbl[[3, 1]],
  tbl[[3, 2]], tbl[[4, 1]], tbl[[4, 2]], tbl[[-1, 1]], GG}, {v1,
  0, 1}, PlotStyle -> {Black, Black, Darker[Blue], Darker[Blue],
   Darker[Red], Darker[Red], Darker[Red], Darker[Red], Darker[Red],
   Darker[Red], Darker[Red], Darker[Green], Darker[Green]}, BaseStyle -> Large]
```

Summary

The simple means have importance as approximations due to their simplicity and bounding character, but they are somewhat imprecise for homogenization tasks with large phase contrasts. Tighter bounds require much more mathematical effort.

Moreover, they cannot account for the geometric arrangement of the phases. Hence, homogenizing isotropic phases with mixture rules that involve only volume fractions, one obtains always effective isotropic materials, even when the microstructure is fibrous or stratified. To obtain the anisotropy due to an anisotropic phase arrangement, one needs more complex methods, which will be discussed in the following chapters.

7 Basic solutions

Basic solutions are the solutions to simple boundary value problems. Well-known examples include the stress field near crack tips from fracture mechanics (see, e. g., Gross and Seelig, 2011), stresses and strains due to an indentation force on an elastic half-space (Boussinesq, 1885[1]-solution), stress concentrations around a hole in a plate (Kirsch, 1898[2]-solution) and Eshelby's solution for an ellipsoidal domain of homogeneous eigenstrains (Eshelby, 1957[3]). A first approach to account for the phase arrangement is the superposition of basic solutions. Often, the Eshelby solution (see Section 9) and the laminate solution (detailed below) are used. An exhaustive presentation can be found in Chapter 9 in Milton (2002).

Another class of results is obtained when the phases' material parameters are fine-tuned without fixing the exact geometry of the microstructure. For example, Hill (1964) found the exact effective moduli for fibrous microstructures of transversely isotropic phases with equal transverse rigidities. Many other special cases exist, some of which are collected in Milton (2002) in Section 5.

We cannot reproduce all basic solutions here but derive at least the laminate solution as the simplest one. This also allows the introduction of the concept of concentration tensors.

7.1 The laminate solution for linear elastic materials

Laminates are quite simple, since the homogeneity of the fields in the phases follows from the homogeneity of the interface orientation. With the jump conditions, one can derive expressions for the effective stiffness, even for nonlinear elasticity and large strains (deBotton, 2005) and for plasticity (Glüge, 2016).

We assume a laminate with volume fractions $v^{\pm}$ and stiffnesses $\mathbb{C}^{\pm}$, with the interface normal vector $\mathbf{n}$, see Figure 7.1.

7.1.1 Jump conditions at the interface

Kinematic and dynamic compatibility at the interface require

$$\boldsymbol{\varepsilon}^{+} - \boldsymbol{\varepsilon}^{-} = \operatorname{sym}(\mathbf{n} \otimes \mathbf{a}), \tag{7.1}$$

$$(\boldsymbol{\sigma}^{+} - \boldsymbol{\sigma}^{-}) \cdot \mathbf{n} = \mathbf{o} \qquad \mathbf{t}^{+} = -\mathbf{t}^{-}. \tag{7.2}$$

The latter equation $\mathbf{t}^{+} = -\mathbf{t}^{-}$ is the local equilibrium per area. With $\mathbf{t} = \boldsymbol{\sigma} \cdot \mathbf{n}$ and $\mathbf{n}^{-} = -\mathbf{n}^{+}$ one obtains $(\boldsymbol{\sigma}^{+} - \boldsymbol{\sigma}^{-}) \cdot \mathbf{n} = \mathbf{o}$. Hence, the stress and strain fields may well

1 Joseph Boussinesq, 1842–1929 (https://en.wikipedia.org/wiki/Joseph_Boussinesq).

2 Ernst Gustav Kirsch, 1841–1901 (https://en.wikipedia.org/wiki/Ernst_Gustav_Kirsch).

3 John Douglas Eshelby, 1916–1981 (https://en.wikipedia.org/wiki/John_D._Eshelby).

https://doi.org/10.1515/9783110793529-007

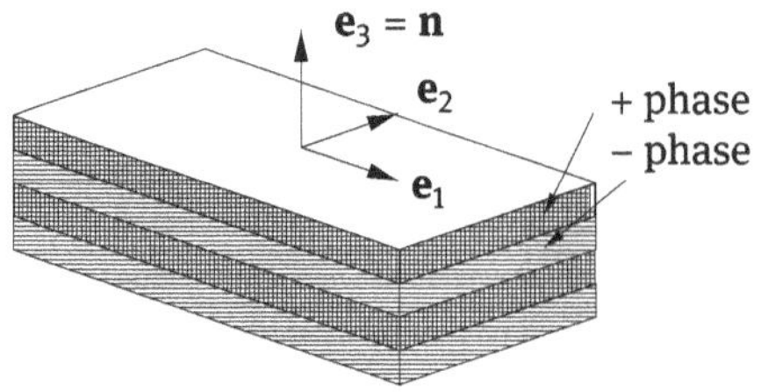

Figure 7.1: Laminate.

have jumps subjected to the constraints that the layers can be joined at the interface. The stress jump needs to be perpendicular to $\mathbf{n}$, and $\boldsymbol{\varepsilon}$ can only have a jump parallel to $\mathbf{n}$. While $\mathbf{t}$ can be decomposed easily into parts parallel and perpendicular to $\mathbf{n}$, the jump of $\boldsymbol{\varepsilon}$ is best written with the unknown quantity $\mathbf{a}$ in $\operatorname{sym}(\mathbf{n} \otimes \mathbf{a})$. Here, we choose $\mathbf{n} = \mathbf{e}_3$. One can decompose $\boldsymbol{\sigma}$ and $\boldsymbol{\varepsilon}$ into the continuous and discontinuous parts,

$$\boldsymbol{\varepsilon}^{\pm} = \underbrace{\varepsilon_1 \mathbf{E}_1 + \varepsilon_2 \mathbf{E}_2 + \varepsilon_4 \mathbf{E}_4}_{\boldsymbol{\varepsilon}_{\text{cont}}} + \underbrace{\varepsilon_3^{\pm} \mathbf{E}_3 + \varepsilon_5^{\pm} \mathbf{E}_5 + \varepsilon_6^{\pm} \mathbf{E}_6}_{\boldsymbol{\varepsilon}^{\pm}_{\text{discont}}}, \tag{7.3}$$

$$\boldsymbol{\sigma}^{\pm} = \underbrace{\sigma_1^{\pm} \mathbf{E}_1 + \sigma_2^{\pm} \mathbf{E}_2 + \sigma_4^{\pm} \mathbf{E}_4}_{\boldsymbol{\sigma}^{\pm}_{\text{discont}}} + \underbrace{\sigma_3 \mathbf{E}_3 + \sigma_5 \mathbf{E}_5 + \sigma_6 \mathbf{E}_6}_{\boldsymbol{\sigma}_{\text{cont}}}. \tag{7.4}$$

In block matrix form, we have

$$\boldsymbol{\varepsilon}^{\pm} = \begin{bmatrix} \varepsilon_{11} & \varepsilon_{12} & \varepsilon_{13}^{\pm} \\ \varepsilon_{12} & \varepsilon_{22} & \varepsilon_{23}^{\pm} \\ \varepsilon_{13}^{\pm} & \varepsilon_{23}^{\pm} & \varepsilon_{33}^{\pm} \end{bmatrix} \mathbf{e}_i \otimes \mathbf{e}_j, \quad \boldsymbol{\sigma}^{\pm} = \begin{bmatrix} \sigma_{11}^{\pm} & \sigma_{12}^{\pm} & \sigma_{13} \\ \sigma_{12}^{\pm} & \sigma_{22}^{\pm} & \sigma_{23} \\ \sigma_{13} & \sigma_{23} & \sigma_{33} \end{bmatrix} \mathbf{e}_i \otimes \mathbf{e}_j, \tag{7.5}$$

where the gray shaded components are continuous. The continuous and discontinuous parts form orthogonal subspaces of the symmetric second-order tensors, w. r. t. the scalar product (Laws, 1975),

$$\boldsymbol{\sigma}^{\pm}_{\text{discont}} : \boldsymbol{\varepsilon}^{\pm}_{\text{discont}} = 0, \tag{7.6}$$

$$\boldsymbol{\sigma}_{\text{cont}} : \boldsymbol{\varepsilon}_{\text{cont}} = 0. \tag{7.7}$$

One can easily define fourth-order projection tensors that filter the continuous and discontinuous parts of $\boldsymbol{\varepsilon}$ and $\boldsymbol{\sigma}$; see, e. g., Hill (1983), He and Feng (2012).

7.1.2 Mixture rule

We assume homogeneity of the fields inside the layers. This piecewise homogeneity satisfies the equilibrium condition $\boldsymbol{\sigma} \cdot \nabla = \mathbf{o}$ trivially. For the effective quantities, we have therefore

$$\bar{\boldsymbol{\sigma}} = v^{-} \boldsymbol{\sigma}^{-} + v^{+} \boldsymbol{\sigma}^{+}, \tag{7.8}$$

$$\bar{\boldsymbol{\varepsilon}} = v^{-} \boldsymbol{\varepsilon}^{-} + v^{+} \boldsymbol{\varepsilon}^{+}, \tag{7.9}$$

with $v^+ + v^- = 1$. This holds, of course, only in the absence of strain localizations. Summarized in a very simplified way, this is the case as long as a strain increment is accompanied by a sufficiently large stress increment, as is the case for elasticity and plasticity with sufficient hardening. This leads us to material laws.

7.1.3 Material laws

We assume anisotropic linear elasticity,

$$\boldsymbol{\sigma}^{\pm} = \mathbb{C}^{\pm} : \boldsymbol{\varepsilon}^{\pm}. \tag{7.10}$$

7.1.4 The effective laminate stiffness

We have 12 strain components in $\boldsymbol{\varepsilon}^{\pm}$ as independent components, which give the stresses $\boldsymbol{\sigma}^{\pm}$ by Hooke's law. The two jump conditions with three equations each allow for a reduction of the independent variables to 6. Due to the linearity of all involved equations, this leads to a linear system of size six, which can be brought into the form

$$\overline{\boldsymbol{\sigma}} = \mathbb{C}^{*} : \overline{\boldsymbol{\varepsilon}}. \tag{7.11}$$

One can then identify $\mathbb{C}^*$. For the derivation, all algebraic manipulations need to be symmetric in the $\pm$-quantities since both quantities are interchangeable.

We begin by inserting the material laws into the jump condition of the stresses,

$$(\boldsymbol{\sigma}^{+} - \boldsymbol{\sigma}^{-}) \cdot \mathbf{n} = (\mathbb{C}^{+} : \boldsymbol{\varepsilon}^{+} - \mathbb{C}^{-} : \boldsymbol{\varepsilon}^{-}) \cdot \mathbf{n}. \tag{7.12}$$

We can now eliminate $\boldsymbol{\varepsilon}^+$ and $\boldsymbol{\varepsilon}^-$ by using the jump conditions of the strains (equation (7.1)),

$$(\mathbb{C}^{+} : \operatorname{sym}(\mathbf{n} \otimes \mathbf{a}) + \Delta\mathbb{C} : \boldsymbol{\varepsilon}^{-}) \cdot \mathbf{n} = \mathbf{o}, \tag{7.13}$$

$$(\mathbb{C}^{-} : \operatorname{sym}(\mathbf{n} \otimes \mathbf{a}) + \Delta\mathbb{C} : \boldsymbol{\varepsilon}^{+}) \cdot \mathbf{n} = \mathbf{o}, \tag{7.14}$$

with the abbreviation $\Delta\mathbb{C} = \mathbb{C}^+ - \mathbb{C}^-$. Because of the symmetrizing effect of $\mathbb{C}$, we can drop the explicit symmetrization. These equations can be rewritten as

$$\mathbf{A}^{+} \cdot \mathbf{a} = -(\Delta\mathbb{C} : \boldsymbol{\varepsilon}^{-}) \cdot \mathbf{n}, \tag{7.15}$$

$$\mathbf{A}^{-} \cdot \mathbf{a} = -(\Delta\mathbb{C} : \boldsymbol{\varepsilon}^{+}) \cdot \mathbf{n}. \tag{7.16}$$

The abbreviation $\mathbf{A}^{\pm} = \mathbf{n} \cdot \mathbb{C}^{\pm} \cdot \mathbf{n}$ is referred to as the acoustic tensor. We can multiply these equations by $v^{\pm}$ and sum them up, such that we can eliminate the local strains by $v^+\boldsymbol{\varepsilon}^+ + v^-\boldsymbol{\varepsilon}^- = \overline{\boldsymbol{\varepsilon}}$,

$$\left(v^- \mathbf{A}^+ + v^+ \mathbf{A}^-\right) \cdot \mathbf{a} = -(\Delta\mathbb{C} : \overline{\boldsymbol{\varepsilon}}) \cdot \mathbf{n}. \tag{7.17}$$

With the abbreviations,

$$\mathbf{Z} = \left(v^- \mathbf{A}^+ + v^+ \mathbf{A}^-\right)^{-1}, \tag{7.18}$$

$$\mathbb{Z} = \mathbf{n} \otimes \mathbf{Z} \otimes \mathbf{n} \tag{7.19}$$

we can write $\mathbf{n} \otimes \mathbf{a}$ compact as

$$\mathbf{n} \otimes \mathbf{a} = -\mathbb{Z} : \Delta\mathbb{C} : \overline{\boldsymbol{\varepsilon}}. \tag{7.20}$$

Next, we generate the effective stresses by using the mixture rules for the stresses,

$$\overline{\boldsymbol{\sigma}} = v^+ \boldsymbol{\sigma}^+ + v^- \boldsymbol{\sigma}^- \tag{7.21}$$

$$= v^+ \mathbb{C}^+ : \boldsymbol{\varepsilon}^+ + v^- \mathbb{C}^- : \boldsymbol{\varepsilon}^-. \tag{7.22}$$

We can again use the jump conditions to replace either $\boldsymbol{\varepsilon}^+$ or $\boldsymbol{\varepsilon}^-$,

$$\overline{\boldsymbol{\sigma}} = \left(v^+ \mathbb{C}^+ + v^- \mathbb{C}^-\right) : \boldsymbol{\varepsilon}^+ - v^- \mathbb{C}^- : \operatorname{sym}(\mathbf{n} \otimes \mathbf{a}), \tag{7.23}$$

$$\overline{\boldsymbol{\sigma}} = \underbrace{\left(v^+ \mathbb{C}^+ + v^- \mathbb{C}^-\right)}_{\overline{\mathbb{C}}} : \boldsymbol{\varepsilon}^- + v^+ \mathbb{C}^+ : \operatorname{sym}(\mathbf{n} \otimes \mathbf{a}). \tag{7.24}$$

Combining $\boldsymbol{\varepsilon}^\pm$ linearly with $v^\pm$ to $\overline{\boldsymbol{\varepsilon}}$, using $v^+ + v^- = 1$ and replacing $\mathbf{n} \otimes \mathbf{a}$ by equation (7.20) one finds finally

$$\overline{\boldsymbol{\sigma}} = \mathbb{C}^* : \overline{\boldsymbol{\varepsilon}} \quad \text{with } \mathbb{C}^* = \overline{\mathbb{C}} - v^+ v^- \Delta\mathbb{C} : \mathbb{Z} : \Delta\mathbb{C}. \tag{7.25}$$

Here, we can identify the effective laminate stiffness. It passes all tests of plausibility: It is invariant under a change of the ±-indexing, has all expected symmetries, and gives the stiffness of the pure phases $\mathbb{C}^\pm$ for the limits $v^\pm \to 1$ or when $\Delta\mathbb{C} = \mathbb{O}$. Moreover, the eigenvalues of $\mathbb{C}^*$ are smaller or equal to the eigenvalues of the Voigt mean $\overline{\mathbb{C}}$. This follows from the positivity of the volume fractions, the quadratic appearance of $\Delta\mathbb{C}$ and the positive definiteness of $\mathbb{C}^\pm$, which is why also $\mathbf{A}^\pm$ and the symmetrized $\mathbb{Z}$ are positive definite. Below this is specified for isotropic phases.

Isotropic phases

For isotropic phases with stiffnesses according to equation (2.21) with the Lamé constants $\lambda^\pm$ and $\mu^\pm$ (equation (2.38)) the components C^*_{ij} of the effective laminate stiffness are

$$\begin{bmatrix} \overline{C} - \frac{v^-v^+(\lambda^- - \lambda^+)^2}{C^+v^- + C^-v^+} & \overline{\lambda} - \frac{v^-v^+(\lambda^- - \lambda^+)^2}{C^+v^- + C^-v^+} & \frac{\lambda^-\lambda^+ - 2\lambda^-\mu^+v^- + 2\lambda^+\mu^-v^+}{C^+v^- + C^-v^+} & 0 & 0 & 0 \\ & \overline{C} - \frac{v^-v^+(\lambda^- - \lambda^+)^2}{C^+v^- + C^-v^+} & \frac{\lambda^-(\lambda^+ - 2\mu^+v^-) + 2\lambda^+\mu^-v^+}{C^+v^- + C^-v^+} & 0 & 0 & 0 \\ & & \frac{C^-C^+}{C^+v^- + C^-v^+} & 0 & 0 & 0 \\ & & & 2\overline{\mu} & 0 & 0 \\ & & & & \frac{2\mu^-\mu^+}{\mu^+v^- + \mu^-v^+} & 0 \\ \text{sym} & & & & & \frac{2\mu^-\mu^+}{\mu^+v^- + \mu^-v^+} \end{bmatrix} \tag{7.26}$$

w. r. t. the normalized Voigt–Mandel notation, with the abbreviations

$$C^{\pm} = \lambda^{\pm} + 2\mu^{\pm} \quad \text{and} \quad \overline{X} = v^+X^+ + v^-X^-. \tag{7.27}$$

For the shear modulus μ, we obtain in the direction of the laminate normal the Reuss mean (C^*_{55} and C^*_{66}) and inside the laminate plane the Voigt mean (C^*_{44}) of the phases' shear moduli $\mu^{\pm}$. The result has been summarized with the aid of Listing 7.1 by inserting the explicit symbolic solution for $\mathbb{C}^*$ (equation (7.25)). But due to the linearity of all involved equations, it is also possible to let a computer algebra system solve for $\mathbb{C}^*$, as demonstrated in Listing 7.2.

Listing 7.1: Explicit calculation (https://gitlab.com/gluegerainer/listings-homogenization-methods/-/blob/main/Listing-07_en.nb) of the laminate stiffness from the symbolic expression (equation (7.25)) with evaluation for isotropic phases.

```
Remove["Global`*"]
(* Stiffness of a laminate of isotropic phases with normal direction e3, p=plus, m=minus *)
id = IdentityMatrix[3]; (* Identity matrix *)
n = {0, 0, 1}; (* e3 is the normal vector *)
id4sym = 1/2 Table[id[[i, k]] id[[j, l]]+id[[i, l]] id[[j, k]],{i,1,3},{j,1,3},{k,1,3},{l,1,3}];
      (* Identity on sym. mat. *)
Cp = lambdap Outer[Times, id, id] + 2 mup id4sym; (* Isotropic stiffnesses *)
Cm = lambdam Outer[Times, id, id] + 2 mum id4sym;
Ap = n.Cp.n; (* Acoustic tensors *)
Am = n.Cm.n;
Z = Inverse[vm Ap + vp Am]; (* Intermediate quantities *)
Z4 = Outer[Times, n, Z, n];
DC = Cp - Cm;
CV = vp Cp + vm Cm;
Ceff = FullSimplify[(CV - vp vm TensorContract[TensorProduct[DC, Z4, DC], {{3, 5}, {4, 6}, {7,
      9}, {8, 10}}]), vm + vp == 1];
(* Function for re-indexing as 6x6 matrices with normalized basis *)
to66[arg_] := ( normalization = Table[If[i <= 3 && j <= 3, 1, If[(i > 3 && j > 3 ), 2, Sqrt
     [2]]], {i, 1, 6}, {j, 1, 6}];
                ivek1 = {1, 2, 3, 1, 1, 2};
                ivek2 = {1, 2, 3, 2, 3, 3};
                Table[normalization[[i, j]] arg[[ivek1[[i]], ivek2[[i]], ivek1[[j]], ivek2[[j
     ]]]], {i, 1, 6}, {j, 1, 6}])
ceff = to66[Ceff];
(* Print result *)
MatrixForm[ceff]
```

Listing 7.2: Calculation (https://gitlab.com/gluegerainer/listings-homogenization-methods/-/blob/main/Listing-08_en.nb) of the laminate stiffness from the jump balances, material laws and mixture rules.

```
Remove["Global`*"]
(* Stiffness of a laminate of isotropic phases with normal direction e3, p=plus, m=minus *)
id = IdentityMatrix[3]; (* Identity matrix and identity on sym. mat. *)
id4sym = 1/2 Table[id[[i,k]]id[[j,l]]+id[[i,l]]id[[j,k]],{i,1,3},{j,1,3},{k,1,3},{l,1,3}];
n = {0, 0, 1}; (* e3 is the normal vector *)
Cp = lambdap Outer[Times, id, id] + 2 mup id4sym; (* Isotropic stiffnesses *)
Cm = lambdam Outer[Times, id, id] + 2 mum id4sym;
(* Unknown rank-1-jump of the strains *)
a = {a1, a2, a3};
jump = 1/2 (Outer[Times, a, n] + Outer[Times, n, a]);
(* Plus-strain as independent variable, minus-strain as dependent variable *)
epsp = {{epsp11, epsp12, epsp13}, {epsp12, epsp22, epsp23}, {epsp13, epsp23, epsp33}};
epsm = epsp - jump;
(* Apply material laws *)
sigmap = Table[Sum[Cp[[i, j, k, l]] epsp[[k, l]], {k, 1, 3}, {l, 1, 3}], {i, 1, 3}, {j, 1, 3}];
sigmam = Table[Sum[Cm[[i, j, k, l]] epsm[[k, l]], {k, 1, 3}, {l, 1, 3}], {i, 1, 3}, {j, 1, 3}];
(* Solve jump balance of stresses for a and assign result *)
eqs = Thread[Flatten /@ ({0, 0, 0} == sigmap.n - sigmam.n)];
erg = Solve[eqs, a];
Set @@@ erg[[1]];
(* Replace plus-strains by effective strains *)
eqs = DeleteDuplicates[Thread[Flatten /@
  ({{eps11, eps12, eps13}, {eps12, eps22, eps23}, {eps13, eps23, eps33}} == vp epsp + vm  epsm)
      ]];
erg = Solve[eqs, DeleteDuplicates[Flatten[epsp]]];
Set @@@ erg[[1]];
(* Write effective stresses as function of effective strains *)
sigma = FullSimplify[vp sigmap + vm sigmam, vm + vp == 1];
(* Extract individual components, e.g. by taking the partial derivative, here for C1212 *)
FullSimplify[D[sigma[[1, 2]], eps12]]
```

7.1.5 Concentration tensors

From the above, we can derive concentration tensors that depend in general on the location, which map the effective stresses or strains to the local stresses or strains. We solve the mixture rule (equation (7.9)) for $v^+\boldsymbol{\varepsilon}^+$,

$$v^+\boldsymbol{\varepsilon}^+ = \overline{\boldsymbol{\varepsilon}} - v^-\boldsymbol{\varepsilon}^-, \tag{7.28}$$

replace $\boldsymbol{\varepsilon}^-$ by the jump condition (equation (7.1)),

$$v^+\boldsymbol{\varepsilon}^+ = \overline{\boldsymbol{\varepsilon}} - v^-(\boldsymbol{\varepsilon}^+ - \mathrm{sym}(\mathbf{n}\otimes\mathbf{a})), \tag{7.29}$$

and solve for $\boldsymbol{\varepsilon}^+$ (with $v^+ + v^- = 1$),

$$\boldsymbol{\varepsilon}^+ = \overline{\boldsymbol{\varepsilon}} + v^-\ \mathrm{sym}(\mathbf{n}\otimes\mathbf{a}). \tag{7.30}$$

This can be summarized further by using the intermediate quantity $\mathbf{n}\otimes\mathbf{a}$ from equation (7.20)

$$\boldsymbol{\varepsilon}^{+} = \underbrace{\left(\mathbb{I} - v^{-}\mathbb{I} : \mathbb{Z} : \Delta\mathbb{C}\right)}_{\mathbb{K}^{+}} : \overline{\boldsymbol{\varepsilon}}, \tag{7.31}$$

where symmetrization is included in the identity on symmetric tensors $\mathbb{I}$. We summarize the linear mapping from $\overline{\boldsymbol{\varepsilon}}$ to $\boldsymbol{\varepsilon}^{+}$ as the concentration tensor $\mathbb{K}^{+}$. Analogously, we have $\mathbb{K}^{-} = \mathbb{I} + v^{+}\mathbb{I} : \mathbb{Z} : \Delta\mathbb{C}$. The concentration tensors for the stresses $\boldsymbol{\sigma}^{\pm} = \mathbb{L}^{\pm} : \overline{\boldsymbol{\sigma}}$ are obtained by extending the $\mathbb{K}^{\pm}$ tensors with the stiffnesses,

$$\mathbb{L}^{\pm} = \mathbb{C}^{\pm} : \mathbb{K}^{\pm} : \mathbb{C}^{*^{-1}}. \tag{7.32}$$

The $\mathbb{K}^{\pm}$-and $\mathbb{L}^{\pm}$ tensors are also termed as polarization tensors. They are not symmetric. Because of

$$\overline{\boldsymbol{\varepsilon}} = \frac{1}{V}\int_{\Omega} \boldsymbol{\varepsilon}(\mathbf{x})\,\mathrm{d}v \tag{7.33}$$

$$= \frac{1}{V}\int_{\Omega} \mathbb{K}(\mathbf{x})\,\mathrm{d}v : \overline{\boldsymbol{\varepsilon}} \tag{7.34}$$

$$= \mathbb{I} : \overline{\boldsymbol{\varepsilon}} \tag{7.35}$$

we have

$$\frac{1}{V}\int_{\Omega} \mathbb{K}(\mathbf{x})\,\mathrm{d}v = \mathbb{I}, \tag{7.36}$$

and likewise for the stresses

$$\frac{1}{V}\int_{\Omega} \mathbb{L}(\mathbf{x})\,\mathrm{d}v = \mathbb{I}, \tag{7.37}$$

i. e., the concentration tensors are, in this sense, normalized.

7.1.6 The laminate solution with eigenstrains

The derivation in Section 7.1 can be repeated with additional eigenstrains in the material laws of the individual phases

$$\boldsymbol{\sigma}^{\pm} = \mathbb{C}^{\pm} : (\boldsymbol{\varepsilon}^{\pm} - \boldsymbol{\varepsilon}^{\pm}_{\mathrm{eig}}). \tag{7.38}$$

This is interesting, e. g., when inhomogeneous thermal expansion is included, or a lattice mismatch between crystalline phases, as may occur in epitaxy. The derivation is analogous to what has been presented above, only that additional terms need to be dragged along. The result is

$$\overline{\boldsymbol{\sigma}} = \mathbb{C}^{*} \cdot\cdot\, \overline{\boldsymbol{\varepsilon}} - \overline{\boldsymbol{\sigma}}_{\mathrm{eig}} + v^{+}v^{-}\Delta\mathbb{C} \cdot\cdot\, \mathbb{Z} \cdot\cdot\, \Delta\boldsymbol{\sigma}_{\mathrm{eig}}, \tag{7.39}$$

with

$$\boldsymbol{\sigma}_{\mathrm{eig}}^{\pm} = \mathbb{C}^{\pm} \cdot\cdot \boldsymbol{\varepsilon}_{\mathrm{eig}}^{\pm}, \tag{7.40}$$

$$\overline{\boldsymbol{\sigma}}_{\mathrm{eig}} = v^{+}\boldsymbol{\sigma}_{\mathrm{eig}}^{+} + v^{-}\boldsymbol{\sigma}_{\mathrm{eig}}^{-}, \tag{7.41}$$

$$\Delta\boldsymbol{\sigma}_{\mathrm{eig}} = \boldsymbol{\sigma}_{\mathrm{eig}}^{+} - \boldsymbol{\sigma}_{\mathrm{eig}}^{-}. \tag{7.42}$$

7.1.7 Remarks regarding the laminate solution

Francfort and Murat (1986) have given the laminate solution for isotropic phases with arbitrary spatial dimension D. Further, Torquato (1997) showed that in the limit $D \to \infty$, the Voigt mean of the shear modulus is the exact solution. This can be understood by considering the jump balances. The phases are in a serial arrangement, which gives the Reuss mean for shear in only $D - 1$ directions in plane and the shear plane normal being the laminate normal, giving $D-1$ serial shear deformation modes. Inside the plane, we have $(D - 2)(D - 1)/2$ possible shear deformations, which deform the phases in parallel, which gives the Voigt mean. To see this, just consider the shear strains in the block matrix in equation (7.5). For three dimensions, we have two serial shear modes and one parallel shear mode, but for higher D, the number of parallel shear modes grows quadratically in D, which quickly swamps the linearly in D growing number of serial shear modes.

With this observation, one could argue that other exponents than $m = -1$ (Reuss), $m = 0$ (geometric mean) and $m = 1$ (Voigt) are interesting in the generalized mean (equation (6.25)). For example, it would be reasonable to use $m = -1/3$ as an equidistant 2/3–1/3-partition of the interval $m = -1 \ldots 1$ since we have with $D = 3$ two serial shear modes and one parallel shear mode; see the lower right 3×3-matrix in equation (7.26). This does not hold for the compression modulus, which contains the dimension D in the case of isotropy, e. g., $3K$ for $D = 3$.

8 Reformulations of the homogenization problem: The eigenstrain problem, the polarization problem and influence tensors

8.1 Difference problem

The differential equation (DE) of the homogenization boundary value problem (the RVE problem) is a homogeneous (zero right-hand side) partial DE with variable coefficients (location-dependent material properties). The goal of the following manipulations is to rewrite this problem as a DE with constant coefficients and a nonzero right-hand side, since for such DE many more solution methods are available.

This is achieved by considering a homogeneous comparison material, the difference of which to the microstructured material is packed into the right-hand side. Consider an RVE problem with linear displacement boundary conditions and inhomogeneous stiffness $\mathbb{C}(\mathbf{x})$,

$$\boldsymbol{\sigma}(\mathbf{x}) \cdot \nabla = \mathbf{o}, \quad \boldsymbol{\sigma}(\mathbf{x}) = \mathbb{C}(\mathbf{x}) : \boldsymbol{\varepsilon}(\mathbf{x}), \quad \boldsymbol{\varepsilon}(\mathbf{x}) = \operatorname{sym}(\mathbf{u}(\mathbf{x}) \otimes \nabla), \quad \mathbf{u}(\mathbf{x})|_{\partial\Omega} = \bar{\boldsymbol{\varepsilon}} \cdot \mathbf{x}. \tag{8.1}$$

The entire linear operator acting on $\mathbf{u}(\mathbf{x})$ is $(\mathbb{C}(\mathbf{x}) : (\mathbf{u}\otimes\nabla))\cdot\nabla$, where the symmetrization is contained in $\mathbb{C}(\mathbf{x})$. The right-hand side is zero. Consider now a comparison problem with the same boundary conditions and a homogeneous stiffness $\mathbb{C}^0$:

$$\boldsymbol{\sigma}^0 \cdot \nabla = \mathbf{o}, \quad \boldsymbol{\sigma}^0 = \mathbb{C}^0 : \boldsymbol{\varepsilon}^0, \quad \boldsymbol{\varepsilon}^0 = (\mathbf{u}^0 \otimes \nabla + \nabla \otimes \mathbf{u}^0)/2, \quad \mathbf{u}^0(\mathbf{x})|_{\partial\Omega} = \bar{\boldsymbol{\varepsilon}} \cdot \mathbf{x}. \tag{8.2}$$

Due to the overall homogeneity of the boundary conditions and $\mathbb{C}^0$ the solution is $\boldsymbol{\varepsilon}(\mathbf{x}) = \boldsymbol{\varepsilon}^0 = \bar{\boldsymbol{\varepsilon}}$. All fields indexed with 0 are therefore homogeneous. We take the difference $\Delta\boldsymbol{\sigma} = \boldsymbol{\sigma} - \boldsymbol{\sigma}^0$. Since all equations are linear, we obtain the difference problem with zero displacement boundary conditions,

$$\Delta\boldsymbol{\sigma}(\mathbf{x}) \cdot \nabla = \mathbf{o}, \quad \Delta\boldsymbol{\sigma}(\mathbf{x}) = \mathbb{C}(\mathbf{x}) : \boldsymbol{\varepsilon}(\mathbf{x}) - \mathbb{C}^0 : \bar{\boldsymbol{\varepsilon}}, \quad \tilde{\boldsymbol{\varepsilon}}(\mathbf{x}) = \operatorname{sym}(\tilde{\mathbf{u}}(\mathbf{x}) \otimes \nabla), \quad \tilde{\mathbf{u}}(\mathbf{x})|_{\partial\Omega} = \mathbf{o}. \tag{8.3}$$

Our aim is to join the unknown field $\tilde{\boldsymbol{\varepsilon}}(\mathbf{x})$ and the homogeneous stiffness $\mathbb{C}^0$. We replace $\bar{\boldsymbol{\varepsilon}} = \boldsymbol{\varepsilon}(\mathbf{x}) - \tilde{\boldsymbol{\varepsilon}}(\mathbf{x})$ in the last equation,

$$\Delta\boldsymbol{\sigma}(\mathbf{x}) = \mathbb{C}(\mathbf{x}) : \boldsymbol{\varepsilon}(\mathbf{x}) - \mathbb{C}^0 : (\boldsymbol{\varepsilon}(\mathbf{x}) - \tilde{\boldsymbol{\varepsilon}}(\mathbf{x})) \tag{8.4}$$

$$= \mathbb{C}^0 : \tilde{\boldsymbol{\varepsilon}}(\mathbf{x}) + \underbrace{(\mathbb{C}(\mathbf{x}) - \mathbb{C}^0) : \boldsymbol{\varepsilon}(\mathbf{x})}_{\boldsymbol{\tau}(\mathbf{x})}. \tag{8.5}$$

The term $\boldsymbol{\tau}(\mathbf{x})$ is called polarization stresses. If we expand $\boldsymbol{\tau}$ from the left with $\mathbb{C}^0 : \mathbb{C}^{0^{-1}}$: and factor out $\mathbb{C}^0$, we obtain

https://doi.org/10.1515/9783110793529-008

$$\Delta\boldsymbol{\sigma}(\mathbf{x}) = \mathbb{C}^0 : [\tilde{\boldsymbol{\varepsilon}}(\mathbf{x}) + \underbrace{\mathbb{C}^{0^{-1}} : (\mathbb{C}(\mathbf{x}) - \mathbb{C}^0) : \boldsymbol{\varepsilon}(\mathbf{x})}_{-\boldsymbol{\varepsilon}_{\text{eig}}(\mathbf{x})}]. \tag{8.6}$$

We consider equation (8.5) as DE for $\tilde{\mathbf{u}}(\mathbf{x})$ and put the polarization or eigenstrain term to the right-hand side. This gives us the desired inhomogeneous DE with constant coefficients. The rewritten problem allows applying a variety of methods, especially Green's function method is now eligible; see Section 12.8.

Although the unknown function is still contained on the right-hand side, this formulation is useful: One can simply consider the polarization problem as a fixed-point iteration rule for the unknown function (see Section 12.7) or make assumptions regarding the right-hand side to derive approximations to the exact solution.

8.2 Homogeneous eigenstrain problem

For the eigenstrain problem, the assumption of piecewise homogeneous eigenstrains is very useful, since we obtain a piecewise homogeneous DE with constant coefficients,

$$\Delta\boldsymbol{\sigma}(\mathbf{x}) = \mathbb{C}^0 : (\tilde{\boldsymbol{\varepsilon}}(\mathbf{x}) - \boldsymbol{\varepsilon}_{\text{eig}\,i}), \quad \text{in phase } i. \tag{8.7}$$

We assume a two-phase matrix-inclusion structure with the stiffnesses $\mathbb{C}^{\text{M}}$ and $\mathbb{C}^{\text{I}}$ in the following sections. We choose $\mathbb{C}^0 = \mathbb{C}^{\text{M}}$, such that $\boldsymbol{\varepsilon}_{\text{eig}}$ is zero in the matrix phase and nonzero in the inclusion. For ellipsoid eigenstrain regions, the solution is given by Eshelby, namely that $\tilde{\boldsymbol{\varepsilon}}(\mathbf{x})$ is homogeneous inside this region and that it vanishes asymptotically outside the eigenstrain region. Interestingly, this homogeneity of the solution $\tilde{\boldsymbol{\varepsilon}}(\mathbf{x})$ in the ellipsoidal region is found in the result and presumed simultaneously in the right-hand side eigenstrain field. We will later use the Eshelby basic solution to derive estimates for $\mathbb{C}^*$ for matrix-inclusion structures.

8.3 Polarization problem

Another way to reformulate the eigenstrain problem is to isolate $\boldsymbol{\varepsilon}$ on the right-hand side. We move $\mathbb{C}^0 : \tilde{\boldsymbol{\varepsilon}}$ to the left-hand side,

$$\underbrace{\Delta\boldsymbol{\sigma} + \mathbb{C}^0 : \tilde{\boldsymbol{\varepsilon}}}_{\mathbf{P}} = \underbrace{(\mathbb{C} - \mathbb{C}^0)}_{\Delta\mathbb{C}} : \boldsymbol{\varepsilon} = \boldsymbol{\sigma} - \mathbb{C}^0 : \boldsymbol{\varepsilon}. \tag{8.8}$$

The left-hand side is abbreviated as $\mathbf{P}$ and called "polarization," likely due to its first usage for electric charge transport. $\mathbf{P}$ is neither divergence-free like $\boldsymbol{\sigma}$ nor is it rotation-free like $\boldsymbol{\varepsilon}$. Moreover, $\mathbf{P}$ has no immediate physical interpretation. This emphasizes that the polarization problem is not solved directly but used as an auxiliary problem to derive expressions for the effective stiffness. One can, for example, take the volume average on both sides of the latter equation and solve for $\overline{\boldsymbol{\sigma}}$,

$$\overline{\boldsymbol{\sigma}} = \overline{\mathbf{P}} + \mathbb{C}^0 : \overline{\boldsymbol{\varepsilon}}. \tag{8.9}$$

Assuming one knows $\overline{\mathbf{P}}$ as a function of $\overline{\boldsymbol{\varepsilon}}$, one can identify the effective stiffness $\mathbb{C}^*$; see Section 12.6.1.

8.4 Influence tensors, concentration tensors, polarization tensors and localization tensors

We have already seen the location dependent concentration tensors, which map effective quantities to local quantities,

$$\boldsymbol{\varepsilon}(\mathbf{x}) = \mathbb{K}(\mathbf{x}) : \overline{\boldsymbol{\varepsilon}}. \tag{8.10}$$

They are also termed localization tensor, influence tensor or polarization tensor. It is the opposite of the eigenstrain concept: we put the location dependence into the concentration tensor $\mathbb{K}$. Let $\boldsymbol{\varepsilon}(\mathbf{x})$ be the strain field of a RVE problem with displacement boundary conditions, which are related to a prescribed average strain $\langle \boldsymbol{\varepsilon} \rangle = \overline{\boldsymbol{\varepsilon}}$. We can write the inhomogeneous stresses and stiffnesses,

$$\boldsymbol{\sigma}(\mathbf{x}) = \mathbb{C}(\mathbf{x}) : \boldsymbol{\varepsilon}(\mathbf{x}), \tag{8.11}$$

$$\overline{\boldsymbol{\sigma}} = \big\langle \mathbb{C}(\mathbf{x}) : \boldsymbol{\varepsilon}(\mathbf{x}) \big\rangle. \tag{8.12}$$

Using the concentration tensor $\mathbb{K}$, we can rewrite this as

$$\overline{\boldsymbol{\sigma}} = \langle \mathbb{C}(\mathbf{x}) : \mathbb{K}(\mathbf{x}) : \overline{\boldsymbol{\varepsilon}} \rangle, \tag{8.13}$$

$$\overline{\boldsymbol{\sigma}} = \underbrace{\big\langle \mathbb{C}(\mathbf{x}) : \mathbb{K}(\mathbf{x}) \big\rangle}_{\mathbb{C}^*} : \overline{\boldsymbol{\varepsilon}}. \tag{8.14}$$

The concentration tensors are like weight factors when averaging the stiffness. We can and will improve the naïve Voigt–Reuss means by good choices for the concentration tensor fields. It is clear that the $\mathbb{K}(\mathbf{x})$-field has the property

$$\big\langle \mathbb{K}(\mathbf{x}) \big\rangle = \mathbb{I} \tag{8.15}$$

of a weighting. Assuming two phases with homogeneous strains, we can write

$$\mathbb{C}^* = v_{\mathrm{I}} \mathbb{C}^{\mathrm{I}} : \mathbb{K}^{\mathrm{I}} + v_{\mathrm{M}} \mathbb{C}^{\mathrm{M}} : \mathbb{K}^{\mathrm{M}}. \tag{8.16}$$

Because of

$$\mathbb{I} = v_\mathrm{I}\mathbb{K}^\mathrm{I} + v_\mathrm{M}\mathbb{K}^\mathrm{M} \tag{8.17}$$

we can eliminate one of the two concentration tensors $\mathbb{K}^{I/M}$. Since we have homogeneous strains in the inclusion and know the Eshelby tensor $\mathbb{E}$ in the ellipsoidal inclusion (see next section) we eliminate $\mathbb{K}^\mathrm{M}$ by solving the last equation for $\mathbb{K}^\mathrm{M}$ and inserting this in the volume average of the stiffness (equation (8.16)),

$$\mathbb{C}^* = v_\mathrm{I}\mathbb{C}^\mathrm{I} : \mathbb{K}^\mathrm{I} + v_\mathrm{M}\mathbb{C}^\mathrm{M} : \mathbb{K}^\mathrm{M} \leftarrow \mathbb{K}^\mathrm{M} = v_\mathrm{M}^{-1}(\mathbb{I} - v_\mathrm{I}\mathbb{K}^\mathrm{I}), \tag{8.18}$$

$$\mathbb{C}^* = v_\mathrm{I}\mathbb{C}^\mathrm{I} : \mathbb{K}^\mathrm{I} + \mathbb{C}^\mathrm{M} : (\mathbb{I} - v_\mathrm{I}\mathbb{K}^\mathrm{I}), \tag{8.19}$$

$$\mathbb{C}^* = \mathbb{C}^\mathrm{M} + v_\mathrm{I}(\mathbb{C}^\mathrm{I} - \mathbb{C}^\mathrm{M}) : \mathbb{K}^\mathrm{I}. \tag{8.20}$$

The last equation will serve us together with the Eshelby solution for deriving various estimates.

9 Improved estimates based on the Eshelby solution

9.1 Eshelby solution

Eshelby (1957) considered the eigenstrain problem for the special case of an ellipsoid region with homogeneous eigenstrains in a homogeneous, infinite body. He found that

- inside the ellipsoid domain, the strain, and hence, stress field turns out to be homogeneous and
- the overall strain inside the ellipsoidal region depends linearly on the eigenstrain,

$$\tilde{\boldsymbol{\varepsilon}} = \mathbb{E} : \boldsymbol{\varepsilon}_{\text{eig}} \tag{9.1}$$

  with the Eshelby tensor $\mathbb{E}$. Unlike $\mathbb{C}$, $\mathbb{E}$ is in general not symmetric.
- For isotropic phases $\mathbb{C}^0$, one can express $\mathbb{E}$ in terms of integrals. For special semi-axes ratios, these integrals can be solved explicitly. Some details can be found in Mura (1987). In the case of a spherical inclusion and an isotropic comparison stiffness $\mathbb{C}^0$, the Eshelby tensor is isotropic, symmetric and given by

$$\mathbb{E} = \alpha \mathbb{P}_{\text{I1}} + \beta \mathbb{P}_{\text{I2}} \tag{9.2}$$

$$\alpha = \frac{3K^0}{3K^0 + 4G^0} = \frac{1 + \nu^0}{3(1 - \nu^0)} \tag{9.3}$$

$$\beta = \frac{6(K^0 + 2G^0)}{5(3K^0 + 4G^0)} = \frac{2(4 - 5\nu^0)}{15(1 - \nu^0)}, \tag{9.4}$$

  where K^0, G^0 and ν^0 are the compression modulus, shear modulus and Poisson ratio of the isotropic material $\mathbb{C}^0 = 3K^0\mathbb{P}_{\text{I1}} + 2G^0\mathbb{P}_{\text{I2}}$.
- Outside the ellipsoidal region, all fields are inhomogeneous, and decay inversely cubic with the distance from the ellipsoid center.

The derivation of the Eshelby solution is quite lengthy. A comprehensible presentation can be found in Yanase (2019).

9.2 Eshelby's concentration tensor

We now put the pieces of the *eigenstrain concept*, *Eshelby solution* and *concentration tensors* together. Consider an ellipsoidal inclusion with stiffness $\mathbb{C}^{\text{I}}$ in a infinite matrix with stiffness $\mathbb{C}^{\text{M}}$. The far strain field is prescribed as $\bar{\boldsymbol{\varepsilon}}$. Our goal is to determine the Eshelby concentration tensor for the inclusion:

1. We choose $\mathbb{C}^0 = \mathbb{C}^{\text{M}}$ in the eigenstrain concept (equation (8.6)). The eigenstrains in the inclusion become

$$\boldsymbol{\varepsilon}_{\text{eig}} = -\mathbb{C}^{\text{M}^{-1}} : [\mathbb{C}^{\text{I}} - \mathbb{C}^{\text{M}}] : \boldsymbol{\varepsilon}. \tag{9.5}$$

https://doi.org/10.1515/9783110793529-009

Outside the inclusion, the eigenstrains are zero since the squared bracket vanishes.

2. This allows applying the Eshelby solution (equation (9.1)). The strains of the difference problem in the inclusion are

$$\tilde{\boldsymbol{\varepsilon}}^* = \mathbb{E} : \boldsymbol{\varepsilon}_{\text{eig}} \tag{9.6}$$

$$= -\mathbb{E} : \mathbb{C}^{\text{M}^{-1}} : [\mathbb{C}^{\text{I}} - \mathbb{C}^{\text{M}}] : \boldsymbol{\varepsilon}. \tag{9.7}$$

Outside the inclusion we cannot give $\tilde{\boldsymbol{\varepsilon}}^*$.

3. Due to the zero displacement boundary conditions, the mean strains of the difference problem are zero. Thus, $\tilde{\boldsymbol{\varepsilon}}^*$ is the unknown deviation $\tilde{\boldsymbol{\varepsilon}}$ from the effective strain $\bar{\boldsymbol{\varepsilon}}$. We can therefore replace $\tilde{\boldsymbol{\varepsilon}}^* = \boldsymbol{\varepsilon} - \bar{\boldsymbol{\varepsilon}}$,

$$\boldsymbol{\varepsilon} - \bar{\boldsymbol{\varepsilon}} = -\mathbb{E} : \mathbb{C}^{\text{M}^{-1}} : [\mathbb{C}^{\text{I}} - \mathbb{C}^{\text{M}}] : \boldsymbol{\varepsilon}. \tag{9.8}$$

4. We can now identify the concentration tensor for the strains inside the inclusion:

$$\boldsymbol{\varepsilon} = \underbrace{(\mathbb{I} + \mathbb{E} : \mathbb{C}^{\text{M}^{-1}} : [\mathbb{C}^{\text{I}} - \mathbb{C}^{\text{M}}])^{-1}}_{\mathbb{K}_{\text{Eshelby}}} : \bar{\boldsymbol{\varepsilon}}. \tag{9.9}$$

In the following sections, we will make use of

$$3K^* = \mathbb{C}^* :: \mathbb{P}_{\text{I1}} \tag{9.10}$$

$$2G^* = \mathbb{C}^* :: \mathbb{P}_{\text{I2}}/5 \tag{9.11}$$

to access the compression and shear moduli K^* and G^* of a stiffness tetrad $\mathbb{C}^*$.

9.3 Dilute distributions

A first improvement of the Voigt–Reuss estimates is obtained by taking the volume average with the Eshelby concentration tensor $\mathbb{K}_{\text{Eshelby}}$ in the inclusion. The starting point is equation (8.20):

$$\mathbb{C}^{DD} = \mathbb{C}^{\text{M}} + v_{\text{I}}(\mathbb{C}^{\text{I}} - \mathbb{C}^{\text{M}}) : \mathbb{K}_{\text{Eshelby}}. \tag{9.12}$$

This yields an affine linear relation for $\mathbb{C}^{\text{DD}}$ in v_{I}. The DD stands for "dilute distribution," which indicates that this is an approximation only for small volume fractions. It ignores the interactions between inclusions. Let us consider stiff inclusions in a soft matrix, all isotropic. We obtain affine linear functions for K^{DD} and G^{DD} in v_{I},

$$K^{\text{DD}} = K^{\text{M}} + v_{\text{I}} \frac{(K^{\text{I}} - K^{\text{M}})(4G^{\text{M}} + 3K^{\text{M}})}{4G^{\text{M}} + 3K^{\text{I}}} \tag{9.13}$$

$$G^{\mathrm{DD}} = G^{\mathrm{M}} + v_{\mathrm{I}} \frac{5(G^{\mathrm{I}} - G^{\mathrm{M}})(4G^{\mathrm{M}} + 3K^{\mathrm{M}})G^{\mathrm{M}}}{6G^{\mathrm{I}}(2G^{\mathrm{M}} + K^{\mathrm{M}}) + G^{\mathrm{M}}(8G^{\mathrm{M}} + 9K^{\mathrm{M}})}; \tag{9.14}$$

see Figure 9.1 for a plot together with the Voigt–Reuss bounds. The corresponding script is given in Listing 9.1.

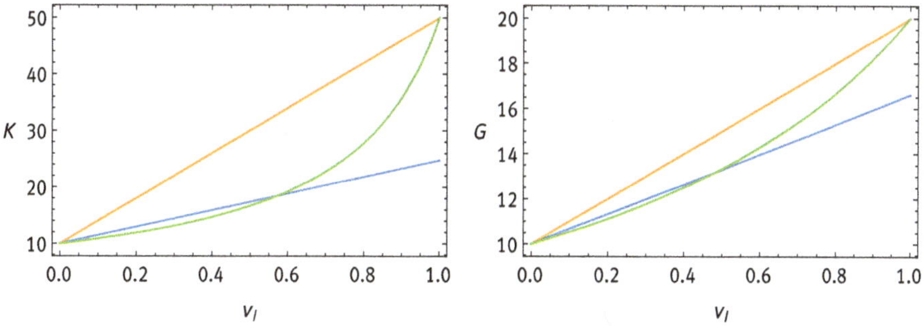

Figure 9.1: Dilute distribution approach: Effective compression modulus K (left) and effective shear modulus G (right) over the inclusion volume fraction v_{I}. One can see that for small v_{I} the estimate lies between the Voigt–Reuss bounds, but for larger v_{I}, the DD estimate becomes so bad that it lies below the Reuss bound.

Listing 9.1: Mathematica notebook (https://gitlab.com/gluegerainer/listings-homogenization-methods/-/blob/main/Listing-09_en.nb) for generating the dilute distribution estimate.

```
Remove["Global`*"];
(* Identity and isotropic projectors *)
id = {1, 1, 1, 0, 0, 0};
P1 = Outer[Times, id, id]/3;
I6 = IdentityMatrix[6];
P2 = I6 - P1;
(* Stiffnesses of inclusion and matrix *)
CI = 3 KI P1 + 2 GI P2;
CM = 3 KM P1 + 2 GM P2;
(* Isotropic Eshelby tensor *)
alpha = 3 KM / (3 KM + 4 GM) ;
beta = 6 (KM + 2 GM) / 5 / (3 KM + 4 GM);
ES = alpha P1 + beta P2;
(* Eshelby concentration tensor *)
K = FullSimplify[Inverse[I6 + ES.Inverse[CM].(CI - CM)]];
(* Effective DD-stiffness *)
Ceff = FullSimplify[CM + vi (CI - CM).K];
(* Project into K and G (note eigenvalue multiplicity of G) *)
KeffDD = FullSimplify[Flatten[Ceff].Flatten[P1]]/3;
GeffDD = FullSimplify[Flatten[Ceff].Flatten[P2]]/2/5;
(* Insert numbers and plot *)
KM = 10; GM = 10; KI = 50; GI = 20;
Plot[{KeffDD, vi KI + (1 - vi) KM, (vi/KI + (1 - vi)/KM)^(-1)},
  {vi, 0, 1.0}, LabelStyle -> Directive[Bold, 20]]
Plot[{GeffDD, vi GI + (1 - vi) GM, (vi/GI + (1 - vi)/GM)^(-1)},
  {vi, 0, 1.0}, LabelStyle -> Directive[Bold, 20]]
```

9.4 Mori–Tanaka

The most obvious improvement of the DD approach is to include the change of the effective matrix stiffness due to the increasing amount of inclusion in the concentration tensor. Instead of $\boldsymbol{\varepsilon}^{\mathrm{I}} = \mathbb{K}_{\text{Eshelby}} : \overline{\boldsymbol{\varepsilon}}$ we use

$$\boldsymbol{\varepsilon}^{\mathrm{I}} = \mathbb{K}_{\text{Eshelby}} : \overline{\boldsymbol{\varepsilon}}^{\mathrm{M}}. \tag{9.15}$$

This can be interpreted as using the effective stiffness as the matrix stiffness. Further, we use

$$\overline{\boldsymbol{\varepsilon}} = v_{\mathrm{I}}\boldsymbol{\varepsilon}^{\mathrm{I}} + v_{\mathrm{M}}\overline{\boldsymbol{\varepsilon}}^{\mathrm{M}}. \tag{9.16}$$

We can now eliminate $\overline{\boldsymbol{\varepsilon}}^{\mathrm{M}}$ in the last two equations and obtain

$$\overline{\boldsymbol{\varepsilon}} = v_{\mathrm{I}}\boldsymbol{\varepsilon}^{\mathrm{I}} + v_{\mathrm{M}}\mathbb{K}_{\text{Eshelby}}^{-1} : \overline{\boldsymbol{\varepsilon}}^{\mathrm{I}}, \tag{9.17}$$

which can be solved for $\boldsymbol{\varepsilon}^{\mathrm{I}}$:

$$\boldsymbol{\varepsilon}^{\mathrm{I}} = \underbrace{\left(v_{\mathrm{I}}\mathbb{I} + v_{\mathrm{M}}\mathbb{K}_{\text{Eshelby}}^{-1}\right)^{-1}}_{\mathbb{K}_{\mathrm{MT}}(v_{\mathrm{I}})} : \overline{\boldsymbol{\varepsilon}}. \tag{9.18}$$

Here, we identify the Mori–Tanaka concentration tensor $\mathbb{K}_{\mathrm{MT}}(v_{\mathrm{I}})$. It is used the same way as in the DD estimate,

$$\mathbb{C}^{*MT} = \mathbb{C}^{\mathrm{M}} + v_{\mathrm{I}}(\mathbb{C}^{\mathrm{I}} - \mathbb{C}^{\mathrm{M}}) : \mathbb{K}_{\mathrm{MT}}(v_{\mathrm{I}}). \tag{9.19}$$

From $\mathbb{C}^{*MT}$, we can extract the compression and shear moduli,

$$K^{\mathrm{MT}} = \frac{12G^{\mathrm{M}}(v_{\mathrm{I}}(K^{\mathrm{I}} - K^{\mathrm{M}}) + K^{\mathrm{M}}) + 9K^{\mathrm{I}}K^{\mathrm{M}}}{3(4G^{\mathrm{M}} + 3(K^{\mathrm{I}}v_{\mathrm{M}} + K^{\mathrm{M}}v_{\mathrm{I}}))} \tag{9.20}$$

$$G^{\mathrm{MT}} = \frac{G^{\mathrm{M}}(G^{\mathrm{I}}(8G^{\mathrm{M}}v_{\mathrm{I}} + 12G^{\mathrm{M}} + 9K^{\mathrm{M}}v_{\mathrm{I}} + 6K^{\mathrm{M}}) + G^{\mathrm{M}}v_{\mathrm{M}}(8G^{\mathrm{M}} + 9K^{\mathrm{M}}))}{G^{\mathrm{M}}(6v_{\mathrm{I}}(2G^{\mathrm{M}} + K^{\mathrm{M}}) + 8G^{\mathrm{M}} + 9K^{\mathrm{M}}) + 6G^{\mathrm{I}}v_{\mathrm{M}}(2G^{\mathrm{M}} + K^{\mathrm{M}})}. \tag{9.21}$$

One can see that $\mathbb{C}^{*MT}$ becomes $\mathbb{C}^{\mathrm{M}}$ and $\mathbb{C}^{\mathrm{I}}$ at $v_{\mathrm{I}} = 0$ and $v_{\mathrm{I}} = 1$. The estimate runs nonlinear in v_i between the Voigt–Reuss bounds; see Figure 9.2. The code additional to Listing 9.1 is given in Listing 9.2.

The Mori–Tanaka estimate is simple but can produce unsymmetric, and hence, nonphysical effective stiffnesses when anisotropic inclusions with different orientations are present; see the introduction in Liu and Huang (2014) for a summary.

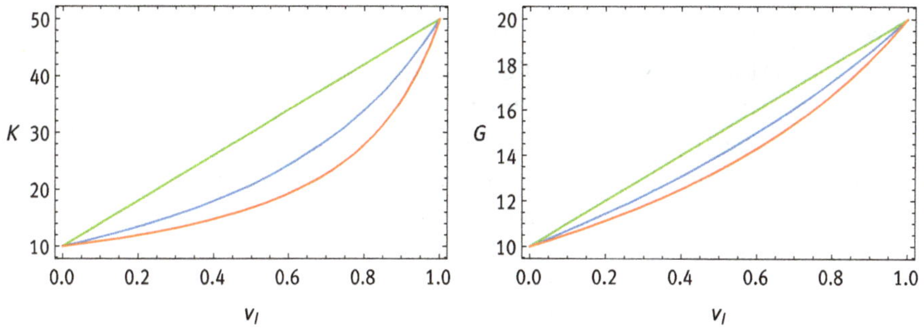

Figure 9.2: Mori–Tanaka approach: Effective compression modulus K (left), effective shear modulus G (right) over v_I. The Mori–Tanaka estimate runs between the Voigt–Reuss bounds.

Listing 9.2: Mathematica notebook (https://gitlab.com/gluegerainer/listings-homogenization-methods/-/blob/main/Listing-09_en.nb) to generate the Mori–Tanaka estimate (additional to Listing 9.1).

```
(* MORI-TANAKA *)
KMT = FullSimplify[Inverse[vi I6 + (1 - vi) Inverse[K]]];
(* Effective MT-stiffness *)
CeffMT = FullSimplify[CM + vi (CI - CM).KMT];
(* Apply projectors on CeffMT to extract K and G (note eigenvalue multiplicity 5 of G) *)
KeffMT = FullSimplify[Flatten[CeffMT].Flatten[P1]]/3;
GeffMT = FullSimplify[Flatten[CeffMT].Flatten[P2]]/2/5;
Plot[{KeffMT, Keff,
  vi KI + (1 - vi) KM, (vi/KI + (1 - vi)/KM)^(-1)}, {vi, 0, 1.0},
 LabelStyle -> Directive[Bold, 20]]
Plot[{GeffMT, Geff,
  vi GI + (1 - vi) GM, (vi/GI + (1 - vi)/GM)^(-1)}, {vi, 0, 1.0},
 LabelStyle -> Directive[Bold, 20]]
```

9.5 Differential scheme

The Mori–Tanaka approach and the dilute distribution approach add a correction term that depends on the inclusion volume fraction to the pure matrix stiffness. Both use different concentration tensors, but in the same linear correction equation (8.20). In the differential scheme, this linear correction equation is used as the linearization of the unknown effective stiffness $\mathbb{C}^*(v_\mathrm{I})$, which is obtained by infinitesimal small, incremental additions of inclusion volume. Thus, to get an estimate for some v_I, one needs to either solve the differential equation or numerically add small quantities of inclusion volume and add up the corrections. The equation

$$\mathbb{C}^* = \mathbb{C}^\mathrm{M} + v_\mathrm{I}(\mathbb{C}^\mathrm{I} - \mathbb{C}^\mathrm{M}) : \mathbb{K}^\mathrm{I} \tag{9.22}$$

is the linearization at $v_\mathrm{I} = 0$. It can be extended to the linearization at arbitrary v_I,

$$\mathbb{C}^\mathrm{DS}(v_\mathrm{I} + \mathrm{d}v_\mathrm{I}) = \mathbb{C}^\mathrm{DS}(v_\mathrm{I}) + \mathrm{d}\tilde{v}_\mathrm{I}(\mathbb{C}^\mathrm{I} - \mathbb{C}^\mathrm{DS}(v_\mathrm{I})) : \mathbb{K}^\mathrm{I}(v_\mathrm{I}), \tag{9.23}$$

but two things need to be accounted for:

– The volume fraction in the above linearization is the matrix volume fraction that is replaced by the inclusion volume fraction at $v_I = 0$. The relative inclusion volume fraction changes $d\tilde{v}_I$ when $v_I > 0$ depends on the content of the inclusion volume fraction, i. e., how many differential steps have been carried out already. Thus, we need to rescale the absolute inclusion volume fraction increment dv_I as if we consider the first increment:

$$d\tilde{v}_I = \frac{dV_I}{V_M} = \frac{dV_I}{V - V_I} = \frac{dv_I}{1 - v_I}. \tag{9.24}$$

One can see that $d\tilde{v}_I$ blows up as v_I approaches 1, i. e., at, say 99 % of inclusion volume fraction, an increment of 1 % is relative to the matrix content that is replaced very large.

– As in the Mori–Tanaka schema, the concentration tensor $\mathbb{K}(v_I)$ contains the stepwise modified matrix stiffness. The stiffness very far from the inclusion is identified as the effective stiffness.

Inserting both into the linearization above gives a tensorial differential equation for $\mathbb{C}^{*DS}$,

$$\mathbb{C}'^{DS}(v_I) = \frac{\mathbb{C}^{DS}(v_I + dv_I) - \mathbb{C}^{DS}(v_I)}{dv_I} \tag{9.25}$$

$$= \frac{1}{1 - v_I}[\mathbb{C}^I - \mathbb{C}^{DS}(v_I)] : \{\mathbb{I} + \mathbb{E}^{DS}(v_I) : \mathbb{C}^{DS^{-1}}(v_I) : [\mathbb{C}^I - \mathbb{C}^{DS}(v_I)]\}^{-1}. \tag{9.26}$$

With isotropic materials and spherical inclusions all involved tensors are isotropic, and hence of the form $\lambda_1\mathbb{P}_{I1} + \lambda_2\mathbb{P}_{I2}$. Therefore, all tensors are coaxial, hence their product is commutative. This allows simplifying the last equation by expanding the first squared bracket with $[\dots]^{-1^{-1}}$ and summarizing with the curly braces $\{\dots\}^{-1}$,

$$\mathbb{C}'^{DS}(v_I) = \frac{1}{1 - v_I}[\mathbb{C}^I - \mathbb{C}^{DS}(v_I)]^{-1^{-1}} : \{\mathbb{I} + \mathbb{E}^{DS}(v_I) : \mathbb{C}^{DS^{-1}}(v_I) : [\mathbb{C}^I - \mathbb{C}^{DS}(v_I)]\}^{-1} \tag{9.27}$$

$$= \frac{1}{1 - v_I}\{[\mathbb{C}^I - \mathbb{C}^{DS}(v_I)]^{-1} + \mathbb{E}^{DS}(v_I) : \mathbb{C}^{DS^{-1}}(v_I)\}^{-1}. \tag{9.28}$$

As before, we obtain the differential equations for $K^{DS}(v_I)$ and $G^{DS}(v_I)$ by taking the scalar product of the tensorial differential equation with the isotropic projectors, $3K^{DS} = \mathbb{C}^{DS} :: \mathbb{P}_{I1}$,

$$3K'^{DS}(v_I) = \frac{1}{1 - v_I}\left[[3K^I - 3K^{DS}(v_I)]^{-1} + \frac{3K^{DS}(v_I)3K^{DS}(v_I)^{-1}}{3K^{DS}(v_I) + 4G^{DS}(v_I)}\right]^{-1} \tag{9.29}$$

$$= \frac{1}{1 - v_I}\left[\frac{1}{3K^I - 3K^{DS}(v_I)} + \frac{1}{3K^{DS}(v_I) + 4G^{DS}(v_I)}\right]^{-1} \tag{9.30}$$

$$= \frac{1}{1 - v_I}\left[\frac{3K^{DS}(v_I) + 4G^{DS}(v_I) + 3K^I - 3K^{DS}(v_I)}{(3K^{DS}(v_I) + 4G^{DS}(v_I)) \cdot (3K^I - 3K^{DS}(v_I))}\right]^{-1} \tag{9.31}$$

$$= \frac{1}{1-v_\mathrm{I}} \frac{(3K^{\mathrm{DS}}(v_\mathrm{I}) + 4G^{\mathrm{DS}}(v_\mathrm{I})) \cdot (3K^\mathrm{I} - 3K^{\mathrm{DS}}(v_\mathrm{I}))}{3K^{\mathrm{DS}}(v_\mathrm{I}) + 4G^{\mathrm{DS}}(v_\mathrm{I}) + 3K^\mathrm{I} - 3K^{\mathrm{DS}}(v_\mathrm{I})} \tag{9.32}$$

$$= \frac{1}{1-v_\mathrm{I}} \frac{(3K^{\mathrm{DS}}(v_\mathrm{I}) + 4G^{\mathrm{DS}}(v_\mathrm{I})) \cdot (3K^\mathrm{I} - 3K^{\mathrm{DS}}(v_\mathrm{I}))}{4G^{\mathrm{DS}}(v_\mathrm{I}) + 3K^\mathrm{I}} \tag{9.33}$$

$$\rightarrow K'^{\mathrm{DS}}(v_\mathrm{I}) = \frac{K^\mathrm{I} - K^{\mathrm{DS}}(v_\mathrm{I})}{1-v_\mathrm{I}} \cdot \frac{3K^{\mathrm{DS}}(v_\mathrm{I}) + 4G^{\mathrm{DS}}(v_\mathrm{I})}{4G^{\mathrm{DS}}(v_\mathrm{I}) + 3K^\mathrm{I}}. \tag{9.34}$$

Analogously, we obtain for $2G^{\mathrm{DS}} = \mathbb{C}^{\mathrm{DS}} :: \mathbb{P}_{\mathrm{I2}}/5$ the resulting scalar differential equation:

$$G'^{\mathrm{DS}}(v_\mathrm{I}) = \frac{5G^{\mathrm{DS}}(v_\mathrm{I})(G^\mathrm{I} - G^{\mathrm{DS}}(v_\mathrm{I}))(4G^{\mathrm{DS}}(v_\mathrm{I}) + 3K^{\mathrm{DS}}(v_\mathrm{I}))}{(1-v_\mathrm{I})[3G^{\mathrm{DS}}(v_\mathrm{I})(4G^\mathrm{I} + 3K^{\mathrm{DS}}(v_\mathrm{I})) + 8G^{\mathrm{DS}}(v_\mathrm{I})^2 + 6G^\mathrm{I}K^{\mathrm{DS}}(v_\mathrm{I})]}. \tag{9.35}$$

Unfortunately, no simple solution can be given to these two nonlinear coupled ordinary first-order DE. The script in Listing 9.3 derives the scalar differential equations for K^{DS} and G^{DS} and numerically integrates our toy problem in the interval $0 \leq v_\mathrm{i} \leq 1$. The matrix stiffness appears as the initial value.

Listing 9.3: Mathematica notebook (https://gitlab.com/gluegerainer/listings-homogenization-methods/-/blob/main/Listing-11_en.nb) to generate the differential scheme estimate.

```
Remove["Global`*"];
(* Differential Scheme *)
(* Identity und isotropic projectors *)
id = {1, 1, 1, 0, 0, 0};
P1 = Outer[Times, id, id]/3;
I6 = IdentityMatrix[6];
P2 = I6 - P1;
(* Stiffnesses *)
KM = KDS[vi];
GM = GDS[vi];
CI = 3 KI P1 + 2 GI P2;
CM = 3 KM P1 + 2 GM P2;
(* Isotropic Eshelby tensor *)
alpha = 3 KM / (3 KM + 4 GM) ;
beta = 6 (KM + 2 GM) / 5 / (3 KM + 4 GM);
ES = alpha P1 + beta P2;
(* Concentration tensor *)
K = Simplify[Inverse[I6 + ES.Inverse[CM].(CI - CM)]];
(* Scalar DE for effective moduli: *)
dglK = KDS'[vi] == FullSimplify[Flatten[((CI - CM).K)].Flatten[P1]/(1 - vi)/3]
dglG = GDS'[vi] == FullSimplify[Flatten[((CI - CM).K)].Flatten[P2]/(1 - vi)/5/2]
(* Fix matrix- and inclusion moduli *)
KM0 = 10; GM0 = 10; KI = 50; GI = 20;
(* Integrate the DE numerically *)
erg = NDSolve[{dglK, dglG, KDS[0] == KM0, GDS[0] == GM0}, {KDS[vi], GDS[vi]}, {vi, 0, 1}]
Plot[{erg[[1, 1, 2]], vi KI + (1 - vi) KM0, (vi/KI + (1 - vi)/KM0)^(-1)}, {vi, 0, 1}]
Plot[{erg[[1, 2, 2]], vi GI + (1 - vi) GM0, (vi/GI + (1 - vi)/GM0)^(-1)}, {vi, 0, 1}]
```

One obtains the DE (9.34) and (9.35). The numerical integration gives the plots in Figure 9.3.

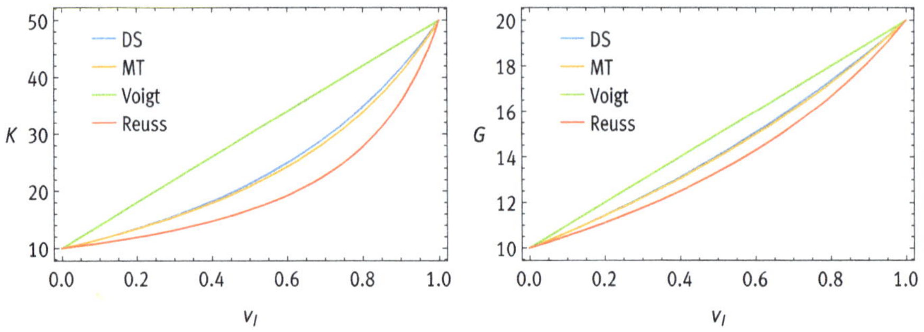

Figure 9.3: Differential scheme: Effective compression modulus K (left) and effective shear modulus G (right) over v_I. Compared to the Mori–Tanaka estimate, the difference is not very large. However, for large phase contrasts, the differential scheme gives results that are much closer to experimental results; see, e. g., Gross and Seelig (2011), Figure 8.22.

9.6 Self-consistent method

The self-consistent method (SC) is a modification of the Mori–Tanaka approach. The MT approach takes the changing matrix stiffness into account by replacing it with the unknown effective stiffness only in the concentration tensor $\mathbb{K}$, without modifying the Eshelby tensor. The self-consistent approach replaces the matrix stiffness also in the Eshelby tensor $\mathbb{E}$. The effective stiffness is then

$$\mathbb{C}^{*SC}(v_\mathrm{I}) = \mathbb{C}^\mathrm{M} + v_\mathrm{I}(\mathbb{C}^\mathrm{I} - \mathbb{C}^\mathrm{M}) : \mathbb{K}_\mathrm{Eshelby}(\mathbb{C}^{*SC}(v_\mathrm{I})) \tag{9.36}$$

with

$$\mathbb{K}_\mathrm{Eshelby}(\mathbb{C}^{*SC}(v_\mathrm{I})) = \left(\mathbb{I} + \mathbb{E}(\mathbb{C}^{*SC}(v_\mathrm{I})) : \mathbb{C}^{*SC}(v_\mathrm{I})^{-1} : [\mathbb{C}^\mathrm{I} - \mathbb{C}^{*SC}(v_\mathrm{I})]\right)^{-1}. \tag{9.37}$$

As a result, one obtains an implicit equation for $\mathbb{C}^{*SC}(v_\mathrm{I})$, which is probably what Mori and Tanaka tried to avoid at the price of replacing the matrix stiffness with the effective stiffness only where it can be handled easily.

For our example of isotropic materials and spherical inclusions, we obtain the coupled equations for K^{SC} and G^{SC} by scalar multiplication with the isotropic projectors, as before. The implementation is given in Listing 9.4.

Listing 9.4: Mathematica notebook (https://gitlab.com/gluegerainer/listings-homogenization-methods/-/blob/main/Listing-12_en.nb) to obtain the self-consistent estimate and the plots in Figure 9.4.

```
Remove["Global`*"];
(* Self-consistent method *)
(* Identity and isotropic projectors *)
id = {1, 1, 1, 0, 0, 0};
P1 = Outer[Times, id, id]/3;
I6 = IdentityMatrix[6];
P2 = I6 - P1;
```

```
(* Stiffnesses *)
CI = 3 KI P1 + 2 GI P2;
CM = 3 KM P1 + 2 GM P2;
(* Approach for unknown effective stiffness *)
CSC = 3 KSC P1 + 2 GSC P2;
(* Eshelby tensor *)
alphaSC = 3 KSC / (3 KSC + 4 GSC) ;
betaSC = 6 (KSC + 2 GSC) / 5 / (3 KSC + 4 GSC);
ESC = alphaSC P1 + betaSC P2;
(* Concentration tensor *)
K4SC = FullSimplify[Inverse[I6 + ESC.Inverse[CSC].(CI - CSC)]];
(* Set up a system of equations of size two *)
eqs = FullSimplify[{
   0 == Flatten[P1].Flatten[CSC - CM - vi*(CI - CM).K4SC],
   0 == Flatten[P2].Flatten[CSC - CM - vi*(CI - CM).K4SC]
   }]
(* Numerical example: stiff inclusions in soft matrix *)
KM = 10; GM = 10; KI = 50; GI = 20;
erg = NSolve[eqs, {GSC, KSC}]
(* The equations are quadratic on the two unknowns. Hence there are  *)
(* 4 solutions, of which one has to select the right one. *)
KeffSC = erg[[4, 2, 2]];
GeffSC = erg[[4, 1, 2]];
Plot[{KeffSC, vi KI + (1 - vi) KM, (vi/KI + (1 - vi)/KM)^(-1)}, {vi, 0, 1.0}]
Plot[{GeffSC, vi GI + (1 - vi) GM, (vi/GI + (1 - vi)/GM)^(-1)}, {vi, 0, 1.0}]
```

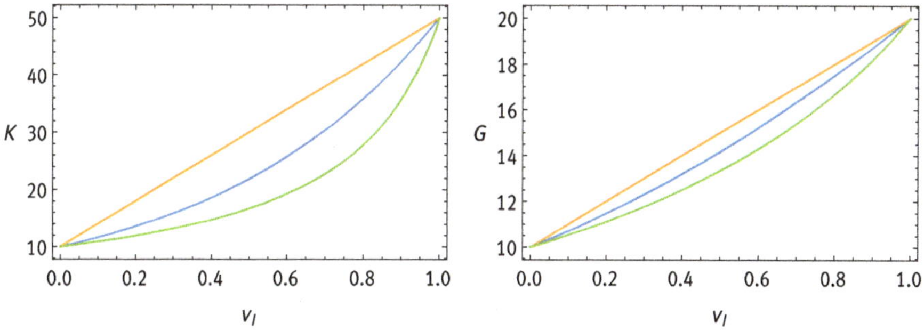

Figure 9.4: Self-consistent estimate: Effective compression modulus K (left) and effective shear modulus G (right) over v_I. The SC estimate runs between the Voigt–Reuss bounds.

We obtain

$$K^{\mathrm{M}} = K^{\mathrm{SC}} + \frac{(-K^{\mathrm{I}} + K^{\mathrm{M}})(4G^{\mathrm{SC}} + 3K^{\mathrm{SC}})v_{\mathrm{I}}}{4G^{\mathrm{SC}} + 3K^{\mathrm{I}}}, \tag{9.38}$$

$$G^{\mathrm{M}} = G^{\mathrm{SC}} + \frac{5(-G^{\mathrm{I}} + G^{\mathrm{M}})G^{\mathrm{SC}}(4G^{\mathrm{SC}} + 3K^{\mathrm{SC}})v_{\mathrm{I}}}{6G^{\mathrm{I}}(2G^{\mathrm{SC}} + K^{\mathrm{SC}}) + G^{\mathrm{SC}}(8G^{\mathrm{SC}} + 9K^{\mathrm{SC}})}. \tag{9.39}$$

Multiplying each equation with the corresponding denominator, factoring out $1 - v_{\mathrm{I}}$ and replacing the latter by v_{M} one finds expressions that are completely symmetric with respect to swapping the denomination of matrix and inclusion,

$$0 = 3K^{\mathrm{I}}K^{\mathrm{M}} - 4G^{\mathrm{SK}}K^{\mathrm{SK}} \tag{9.40}$$
$$+ (4G^{\mathrm{SK}}K^{\mathrm{I}} - 3K^{\mathrm{M}}K^{\mathrm{SK}})v_{\mathrm{I}} \tag{9.41}$$
$$+ (4G^{\mathrm{SK}}K^{\mathrm{M}} - 3K^{\mathrm{I}}K^{\mathrm{SK}})v_{\mathrm{M}} \tag{9.42}$$
$$0 = -12G^{\mathrm{I}}G^{\mathrm{M}}G^{\mathrm{SK}} + 8{G^{\mathrm{SK}}}^3 - 6G^{\mathrm{I}}G^{\mathrm{M}}K^{\mathrm{SK}} + 9{G^{\mathrm{SK}}}^2K^{\mathrm{SK}} \tag{9.43}$$
$$+ (12G^{\mathrm{I}}{G^{\mathrm{SK}}}^2 - 8G^{\mathrm{M}}{G^{\mathrm{SK}}}^2 + 6G^{\mathrm{I}}G^{\mathrm{SK}}K^{\mathrm{SK}} - 9G^{\mathrm{M}}G^{\mathrm{SK}}K^{\mathrm{SK}})v_{\mathrm{M}} \tag{9.44}$$
$$+ (12G^{\mathrm{M}}{G^{\mathrm{SK}}}^2 - 8G^{\mathrm{I}}{G^{\mathrm{SK}}}^2 + 6G^{\mathrm{M}}G^{\mathrm{SK}}K^{\mathrm{SK}} - 9G^{\mathrm{I}}G^{\mathrm{SK}}K^{\mathrm{SK}})v_{\mathrm{I}}. \tag{9.45}$$

This comes as a surprise since it means that both phases are treated equally in the SC approach, although we use the Eshelby solution that clearly distinguishes between matrix and inclusion. One may find this not so self-consistent after all. Nevertheless, the SC approach may be useful for grain structures: A single grain is an inclusion, but there is no pure matrix phase. The SC-approach gives good results only for not too large inclusion volume fractions.

10 Percolation bounds

The above approaches can yield curious results when the phase contrast becomes very large, i. e., when the inclusions become voids (zero stiffness) or rigid (infinite stiffness). In such situations, abrupt changes in the effective material behavior are observed when the inclusions form percolating networks at critical volume fractions. For example, if the inclusions are voids, one may speak of fracture once this critical point is reached, such that the effective stiffness also becomes zero. Percolation theory is concerned with determining such critical volume fractions. It is a purely geometric effect that also plays an important role for diffusion, i. e., at which pore volume fraction the material becomes permeable.

Figure 10.1 is taken from Gross and Seelig (2011). It shows the effective stiffness of a plate with holes that has been determined experimentally, as well as the predictions from different approaches. One can see that only the self-consistent scheme gives a reasonable percolation threshold, which differs nonetheless considerably from the experimental result.

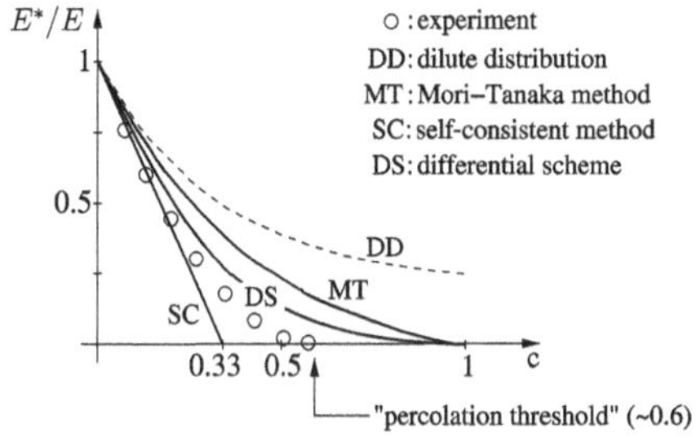

Figure 10.1: Effective Young's modulus of a plate with randomly distributed holes (Gross and Seelig, 2011).

The effect of percolation depends strongly on the difference in the phase's material properties. A nice introduction, including its relation to homogenization, is given in Chen (2008).

10.1 Discrete percolation models

Discrete percolation models consider network-like connections between points, areas, or volumes. It is distinguished between bond and site percolation. For bond percolation, the connection between points is important, e. g., in the case of a lattice in which the lattice points are connected along the grid lines; see Figure 10.2 right. At a critical ratio of active connections to total connections, percolation takes place. In case of site percolation, the presence or absence of elements (e. g., squares on a checkerboard) themselves is considered, and neighboring elements that are present count as connected; see Figure 10.2 left.

https://doi.org/10.1515/9783110793529-010

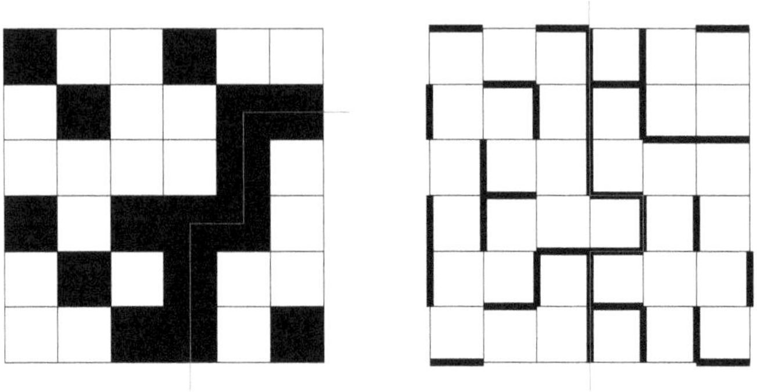

Figure 10.2: Site percolation on the left and bond percolation on the right on a Cartesian grid. Percolation paths are drawn in pink.

Simple expressions for percolation limits can only be given for discrete regular lattices (trigonal, tetragonal, hexagonal); see Chen (2008), Table 2 or https://en.wikipedia.org/wiki/Percolation_threshold for an exhaustive list of percolation limits. Additionally, one can ask for specific percolation directions. Percolation of water through soil generally follows gravity, and fracture is, in a strict sense, only problematic when tensile stresses perpendicular to the crack plane occur.

10.2 Continuous percolation models

For continuous percolation, one can hardly give any simple solutions but needs to resort to numerical experiments with RVEs. In the latter, the inclusions or cracks are randomly dispersed, and percolation is checked numerically. In the 1980s and 1990s, many such studies were conducted, and critical volume fractions for all kinds of inclusions were determined. As an example, we depict here the percolation limits for ellipsoidal inclusions in Figure 10.3. They have been determined numerically by Garboczi (1995). One can see a Padé[1] approximation that has been fitted to the numerical results, which depend on the half-axis ratios of the ellipsoids. One notes a nice, smooth curve and that percolation volume fractions are smaller for half-axis ratios far away from one, i. e., the more ellipsoid the inclusions are. For spherical inclusions, the critical volume fraction is the largest, with approximately 0.3. This is somewhat far from the densest sphere packing of $\pi/\sqrt{18} \approx 0.74$.[2] In the 2D case, the percolation limit for circular holes is approximately 0.6, see Figure 10.1.

1 Henri Padé, 1863–1953 (https://en.wikipedia.org/wiki/Henri_Pade).

2 Kepler's conjecture (Johannes Kepler, 1571–1630 (https://en.wikipedia.org/wiki/Johannes_Kepler)) was proved at the beginning of the 21st century with the aid of computer algebra and formally, without computer-aided math, in 2017.

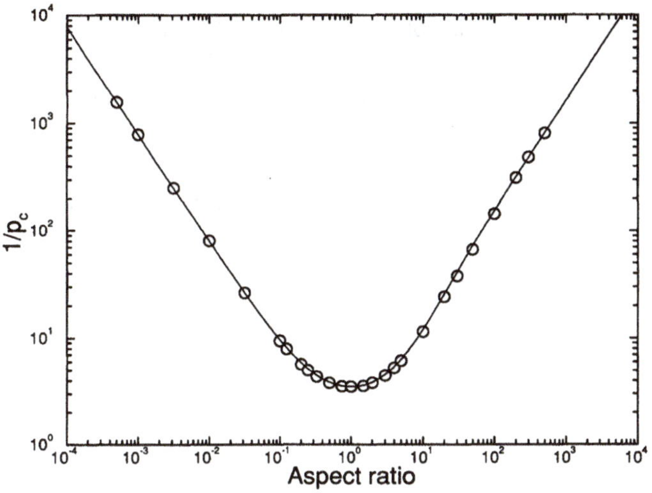

Figure 10.3: From Garboczi (1995): The inverse of the percolation inclusion volume fraction $1/p$ over the half-axis ratio of rotation ellipsoids that are dispersed randomly in a matrix. The line is a Padé approximation of the numerical values.

10.3 Relation between percolation and effective properties

Erroneously, one may be tempted to approximate critical volume fractions by considering largely different phase stiffnesses in the estimates presented in Chapter 9 and identify a point of transition in the effective stiffness. In a logarithmic scaling, this point is characterized by a nearly vertical curve; see Figure 10.4. Except for the self-consistent method, this leads to unrealistic percolation volume fractions at volume fractions of 0 % and 100 %. Due to the symmetric treatment of the phases, we obtain a critical value of 50 % for the self-consistent method. In Figure 10.4, the graph of $G_{SK}(v_I)$ is given with $G_M = 1$, $G_I = 0$, $K_I = 0$ for different K_M. The volume fraction at which $G_{SK} = 0$ is exactly 0.5, which is much bigger than the realistic value of ap-

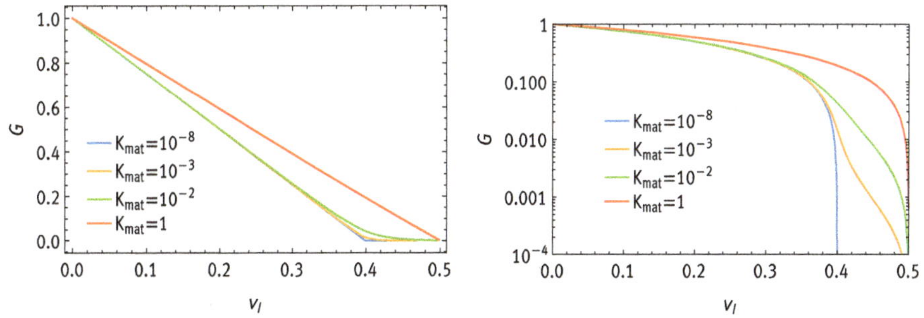

Figure 10.4: Self-consistent estimate with $G_M = 1$, $G_I = 0$, $K_I = 0$ for different K_M.

proximately 0.3. Interestingly, $G_{SK} = 0$ is obtained near $v_I = 0.4$ for smaller K_M. In Andrianov et al. (2018), this is explored in more detail.

We will not dive further into this since it is conceptually not reasonable. Percolation limits can be determined purely geometrically and should not depend on the phase's material properties.

Instead, the percolation limit should be used by homogenization methods to have more informed approaches for the effective properties, as in Chen (2008) (see Figure 15 therein), but not the other way around.

11 Hashin–Shtrikman variational method

We have seen that the Voigt–Reuss bounds can span a rather wide interval. It is, therefore, reasonable to search for tighter bounds. In the following, the method of Hashin and Shtrikman[1] (Hashin and Shtrikman, 1962b, 1963; Watt and Peselnick, 1980) is presented. They construct a functional for the difference problem that

- reduces to the minimum principle of the complementary energy for $\mathbb{C}^0 \to \mathbb{O}$,
- reduces to the minimum principle of the elastic potential for $\mathbb{C}^0 \to \infty$,
- takes the value of the effective strain energy at the solution,
- takes a maximum or a minimum value when $\mathbb{C}^0$ is sufficiently small or large.

Due to the last property, one can derive bounds similar to the ones in Section 6.1, which are tighter than the Voigt–Reuss bounds because

- The value of the functional can be adjusted with $\mathbb{C}^0$.
- The difference problem takes the homogeneously deformed state as the starting point, which is closer to the solution than the undeformed state.
- The Hashin–Shtrikman bounds are derived by approaching the fields inside the RVE as piecewise homogeneous, whereas the Voigt–Reuss bounds presume completely homogeneous strains or stresses.
- Assumptions regarding the microstructure arrangement are used, namely isotropy, unlike for the classical bounds.

The book of Milton (2002) is somewhat exhaustive regarding the different bounds. The following derivation follows the work of Hashin and Shtrikman (1962b) closely but is more exhaustive and, therefore, split into several steps:

1. First, we generalize the minimization principles equation (6.10) and equation (6.16) to the Hashin–Shtrikman functional over the two fields $\tilde{\boldsymbol{\varepsilon}}$ and $\boldsymbol{\tau}$ that are constraint via the solution of the difference problem. The functional contains the comparison stiffness as a free parameter. This is done in the present section and the following Section 11.1.
2. We then show the functional's properties, namely that the solution of the difference problem minimizes or maximizes this functional, depending on the choice of the comparison stiffness. Moreover, the functional's exact solution value corresponds to the effective stored elastic energy, Sections 11.2 to 11.3.
3. Using a Fourier method and presuming isotropy of the microstructure, it is shown that we can replace $\tilde{\boldsymbol{\varepsilon}} = \mathbb{W} : \tilde{\boldsymbol{\tau}}$ with $\mathbb{W}$ depending only on the choice of $\mathbb{C}^0$, Section 12.5.
4. Finally, the functional is mini- or maximized (depending on $\mathbb{C}^0$) over the parameters of a piecewise homogeneous approach for $\boldsymbol{\tau}$, Section 11.5.

1 Shmuel Shtrikman, 1930–2003 (https://en.wikipedia.org/wiki/Shmuel_Shtrikman).

https://doi.org/10.1515/9783110793529-011

For the derivation, we will not denote the location dependence of the fields. It suffices to remember that all quantities without a superscript 0 or a superimposed bar depend on $\mathbf{x}$. Starting point is the difference problem with eigenstrains (Chapter 8), specifically equations (8.5) and (8.3). As an intermediate quantity, the polarization stresses $\boldsymbol{\tau}$ appear,

$$\boldsymbol{\tau} = \Delta\mathbb{C} : \boldsymbol{\varepsilon}, \tag{11.1}$$

with the abbreviation $\Delta\mathbb{C} = \mathbb{C} - \mathbb{C}^0$. They give the difference between the location-dependent inhomogeneous stresses and the stresses in the comparison medium. We now consider $\boldsymbol{\tau}$ and $\tilde{\boldsymbol{\varepsilon}}$ as unknown functions, which are nevertheless not independent of each other. We construct a functional $F(\hat{\boldsymbol{\tau}}, \hat{\tilde{\boldsymbol{\varepsilon}}})$ which has equation (11.1) as the solution:

$$F(\hat{\boldsymbol{\tau}}, \hat{\tilde{\boldsymbol{\varepsilon}}}) = \frac{1}{2V} \int_{\Omega} \bar{\boldsymbol{\varepsilon}} : \mathbb{C}^0 : \bar{\boldsymbol{\varepsilon}} - \hat{\boldsymbol{\tau}} : \Delta\mathbb{C}^{-1} : \hat{\boldsymbol{\tau}} + \hat{\tilde{\boldsymbol{\varepsilon}}} : \hat{\boldsymbol{\tau}} + 2\hat{\boldsymbol{\tau}} : \bar{\boldsymbol{\varepsilon}} \, \mathrm{d}V. \tag{11.2}$$

The fields $\hat{\boldsymbol{\tau}}$ and $\hat{\tilde{\boldsymbol{\varepsilon}}}$ need to satisfy the equilibrium, hence

$$\left(\mathbb{C}^0 : \hat{\tilde{\boldsymbol{\varepsilon}}} + \hat{\boldsymbol{\tau}}\right) \cdot \nabla = \mathbf{o}. \tag{11.3}$$

In the following, we drop the scalar contractions ":," since there is always only one way to summarize the terms to a scalar.

11.1 The solution

We first show that the functional (equation (11.2)) gives the solution to the difference problem (equation (11.1)) upon maximizing or minimizing over $\boldsymbol{\tau}$ and $\tilde{\boldsymbol{\varepsilon}}$ under the constraint that $\boldsymbol{\tau}$ and $\tilde{\boldsymbol{\varepsilon}}$ are in equilibrium.

To do so, we determine the first variation δF by considering the fields $\hat{\boldsymbol{\tau}}$, $\hat{\tilde{\boldsymbol{\varepsilon}}}$ as the solution $\boldsymbol{\tau}$, $\tilde{\boldsymbol{\varepsilon}}$ plus the deviation from the solution $\delta\boldsymbol{\tau}$, $\delta\tilde{\boldsymbol{\varepsilon}}$ scaled by factors α, β,

$$\hat{\boldsymbol{\tau}} = \boldsymbol{\tau} + \alpha\delta\boldsymbol{\tau}, \quad \hat{\tilde{\boldsymbol{\varepsilon}}} = \tilde{\boldsymbol{\varepsilon}} + \beta\delta\tilde{\boldsymbol{\varepsilon}}. \tag{11.4}$$

At the solution $\alpha = 0$ and $\beta = 0$, the differential vanishes, $\mathrm{d}F = 0$,

$$0 = \mathrm{d}F = \frac{\partial F}{\partial \alpha}\mathrm{d}\alpha + \frac{\partial F}{\partial \beta}\mathrm{d}\beta. \tag{11.5}$$

We add up the partial derivatives w. r. t. α and β and then set $\alpha = 0$ and $\beta = 0$,

$$\delta F = \left.\frac{\partial F(\boldsymbol{\tau} + \alpha\delta\boldsymbol{\tau}, \tilde{\boldsymbol{\varepsilon}} + \beta\delta\tilde{\boldsymbol{\varepsilon}})}{\partial \alpha}\right|_{\alpha=0,\beta=0} + \left.\frac{\partial F(\boldsymbol{\tau} + \alpha\delta\boldsymbol{\tau}, \tilde{\boldsymbol{\varepsilon}} + \beta\delta\tilde{\boldsymbol{\varepsilon}})}{\partial \beta}\right|_{\alpha=0,\beta=0}. \tag{11.6}$$

The result is

$$\delta F = \frac{1}{V}\int_{\Omega} -\boldsymbol{\tau}\Delta\mathbb{C}^{-1}\delta\boldsymbol{\tau} + \frac{1}{2}\tilde{\boldsymbol{\varepsilon}}\delta\boldsymbol{\tau} + \bar{\boldsymbol{\varepsilon}}\delta\boldsymbol{\tau} + \frac{1}{2}\boldsymbol{\tau}\delta\tilde{\boldsymbol{\varepsilon}}\,\mathrm{d}V. \tag{11.7}$$

We expand the integrand with $\frac{1}{2}\tilde{\boldsymbol{\varepsilon}}\delta\boldsymbol{\tau} - \frac{1}{2}\tilde{\boldsymbol{\varepsilon}}\delta\boldsymbol{\tau}$ and summarize with $\boldsymbol{\varepsilon} = \tilde{\boldsymbol{\varepsilon}} + \bar{\boldsymbol{\varepsilon}}$ to obtain

$$\delta F = \frac{1}{V}\int_{\Omega} (-\boldsymbol{\tau}\Delta\mathbb{C}^{-1} + \boldsymbol{\varepsilon})\delta\boldsymbol{\tau} - \frac{1}{2}\tilde{\boldsymbol{\varepsilon}}\delta\boldsymbol{\tau} + \frac{1}{2}\boldsymbol{\tau}\delta\tilde{\boldsymbol{\varepsilon}}\,\mathrm{d}V. \tag{11.8}$$

We will show below that the two last summands vanish. Since $\delta\boldsymbol{\tau}$ is nonzero, the bracket must be the solution $\boldsymbol{\tau} = \Delta\mathbb{C}\boldsymbol{\varepsilon}$, which is what we want to demonstrate. We now check the last two summands of the above integral,

$$\delta F = \frac{1}{V}\int_{\Omega} -\frac{1}{2}\tilde{\boldsymbol{\varepsilon}}\delta\boldsymbol{\tau} + \frac{1}{2}\boldsymbol{\tau}\delta\tilde{\boldsymbol{\varepsilon}}\,\mathrm{d}V. \tag{11.9}$$

For this, we need to make use of the constraints of the field $\tilde{\boldsymbol{\varepsilon}}$ and

$$\Delta\boldsymbol{\sigma} = \mathbb{C}^{0}\tilde{\boldsymbol{\varepsilon}} + \boldsymbol{\tau}, \quad \Delta\boldsymbol{\sigma}\cdot\nabla = \mathbf{o}. \tag{11.10}$$

The auxiliary field $\Delta\boldsymbol{\sigma}$ is due to the constraint equation (11.3), which is divergence-free. The same holds for the variation,

$$\delta\Delta\boldsymbol{\sigma} = \mathbb{C}^{0}\delta\tilde{\boldsymbol{\varepsilon}} + \delta\boldsymbol{\tau}, \quad \delta\Delta\boldsymbol{\sigma}\cdot\nabla = \mathbf{o}. \tag{11.11}$$

We solve the equations for $\Delta\boldsymbol{\sigma}$ and $\delta\Delta\boldsymbol{\sigma}$ for $\boldsymbol{\tau}$ and $\delta\boldsymbol{\tau}$ and replace these in equation (11.9),

$$\delta F = \frac{1}{V}\int_{\Omega} -\frac{1}{2}\tilde{\boldsymbol{\varepsilon}}(\delta\Delta\boldsymbol{\sigma} - \mathbb{C}^{0}\delta\tilde{\boldsymbol{\varepsilon}}) + \frac{1}{2}(\Delta\boldsymbol{\sigma} - \mathbb{C}^{0}\tilde{\boldsymbol{\varepsilon}})\delta\tilde{\boldsymbol{\varepsilon}}\,\mathrm{d}V, \tag{11.12}$$

$$= \frac{1}{V}\int_{\Omega} -\frac{1}{2}\tilde{\boldsymbol{\varepsilon}}\delta\Delta\boldsymbol{\sigma} + \frac{1}{2}\tilde{\boldsymbol{\varepsilon}}\mathbb{C}^{0}\delta\tilde{\boldsymbol{\varepsilon}} + \frac{1}{2}\Delta\boldsymbol{\sigma}\delta\tilde{\boldsymbol{\varepsilon}} - \frac{1}{2}\delta\tilde{\boldsymbol{\varepsilon}}\mathbb{C}^{0}\tilde{\boldsymbol{\varepsilon}}\,\mathrm{d}V \tag{11.13}$$

$$= \frac{1}{V}\int_{\Omega} -\frac{1}{2}\tilde{\boldsymbol{\varepsilon}}\delta\Delta\boldsymbol{\sigma} + \frac{1}{2}\Delta\boldsymbol{\sigma}\delta\tilde{\boldsymbol{\varepsilon}}\,\mathrm{d}V, \tag{11.14}$$

where we use the symmetry of $\mathbb{C}^{0}$. We next replace $\tilde{\boldsymbol{\varepsilon}} = \mathrm{sym}(\tilde{\mathbf{u}}\otimes\nabla)$ and $\delta\tilde{\boldsymbol{\varepsilon}} = \mathrm{sym}(\delta\tilde{\mathbf{u}}\otimes\nabla)$ in index notation. We can drop the symmetrization, as the antisymmetric part has no effect on the outcome due to the symmetry of $\Delta\boldsymbol{\sigma}$ and $\delta\Delta\boldsymbol{\sigma}$:

$$\delta F = \frac{1}{V}\int_{\Omega} -\frac{1}{2}\tilde{u}_{i,j}\delta\sigma_{ij} + \frac{1}{2}\sigma_{ij}\delta\tilde{u}_{i,j}\,\mathrm{d}V. \tag{11.15}$$

Carrying out the product rule backwards gives

$$\delta F = \frac{1}{V}\int_{\Omega} -\frac{1}{2}(\tilde{u}_i\delta\sigma_{ij})_{,j} + \frac{1}{2}\tilde{u}_i\delta\sigma_{ij,j} + \frac{1}{2}(\sigma_{ij}\delta\tilde{u}_i)_{,j} - \frac{1}{2}\sigma_{ij,j}\delta\tilde{u}_i \, \mathrm{d}V. \tag{11.16}$$

$\sigma_{ij,j}$ and $\delta\sigma_{ij,j}$ are zero due to $\Delta\boldsymbol{\sigma}$ and $\delta\Delta\boldsymbol{\sigma}$ being divergence-free. We now apply the Gauß–Ostrogradski theorem and convert the volume integral to a surface integral,

$$\delta F = \frac{1}{V}\int_{\partial\Omega} -\frac{1}{2}\tilde{u}_i\delta\Delta\sigma_{ij}n_j + \frac{1}{2}\Delta\sigma_{ij}\delta\tilde{u}_i n_j \, \mathrm{d}A. \tag{11.17}$$

$\tilde{\mathbf{u}}$ is the displacement on the boundary of the difference problem according to equation (8.3), which is by definition zero. Therefore, the entire integral is zero, which shows that the functional takes a stationary value at the solution of the difference problem $\boldsymbol{\tau} = \Delta\mathbb{C} : \boldsymbol{\varepsilon}$.

11.2 The functional's value at the solution

To obtain the value of the Hashin–Shtrikman functional at the solution, we insert the solution $\boldsymbol{\tau} = \Delta\mathbb{C}\boldsymbol{\varepsilon}$ and summarize

$$F = \frac{1}{2V}\int_{\Omega} \bar{\boldsymbol{\varepsilon}}\mathbb{C}^0\bar{\boldsymbol{\varepsilon}} - \boldsymbol{\tau}\Delta\mathbb{C}^{-1}\boldsymbol{\tau} + \tilde{\boldsymbol{\varepsilon}}\boldsymbol{\tau} + 2\boldsymbol{\tau}\bar{\boldsymbol{\varepsilon}} \, \mathrm{d}V \leftarrow \boldsymbol{\tau} = \Delta\mathbb{C}\boldsymbol{\varepsilon} \tag{11.18}$$

$$= \frac{1}{2V}\int_{\Omega} \bar{\boldsymbol{\varepsilon}}\mathbb{C}^0\bar{\boldsymbol{\varepsilon}} - \boldsymbol{\varepsilon}\Delta\mathbb{C}\boldsymbol{\varepsilon} + \tilde{\boldsymbol{\varepsilon}}\Delta\mathbb{C}\boldsymbol{\varepsilon} + 2\boldsymbol{\varepsilon}\Delta\mathbb{C}\bar{\boldsymbol{\varepsilon}} \, \mathrm{d}V. \tag{11.19}$$

We replace $\tilde{\boldsymbol{\varepsilon}} = \boldsymbol{\varepsilon} - \bar{\boldsymbol{\varepsilon}}$ and summarize further,

$$F = \frac{1}{2V}\int_{\Omega} \bar{\boldsymbol{\varepsilon}}\mathbb{C}^0\bar{\boldsymbol{\varepsilon}} - \boldsymbol{\varepsilon}\Delta\mathbb{C}\boldsymbol{\varepsilon} + \boldsymbol{\varepsilon}\Delta\mathbb{C}\boldsymbol{\varepsilon} - \bar{\boldsymbol{\varepsilon}}\Delta\mathbb{C}\boldsymbol{\varepsilon} + 2\boldsymbol{\varepsilon}\Delta\mathbb{C}\bar{\boldsymbol{\varepsilon}} \, \mathrm{d}V \tag{11.20}$$

$$= \frac{1}{2V}\int_{\Omega} \bar{\boldsymbol{\varepsilon}}\mathbb{C}^0\bar{\boldsymbol{\varepsilon}} + \bar{\boldsymbol{\varepsilon}}\Delta\mathbb{C}\boldsymbol{\varepsilon} \, \mathrm{d}V. \tag{11.21}$$

Using $\Delta\mathbb{C} = \mathbb{C} - \mathbb{C}^0$, we finally obtain

$$F = \frac{1}{2V}\int_{\Omega} \bar{\boldsymbol{\varepsilon}}\mathbb{C}^0\bar{\boldsymbol{\varepsilon}} + \bar{\boldsymbol{\varepsilon}}\mathbb{C}\boldsymbol{\varepsilon} - \bar{\boldsymbol{\varepsilon}}\mathbb{C}^0\boldsymbol{\varepsilon} \, \mathrm{d}V. \tag{11.22}$$

Integration of the first and last summand gives $\pm 1/2\bar{\boldsymbol{\varepsilon}}\mathbb{C}^0\bar{\boldsymbol{\varepsilon}}$, which cancel each other out. The first summand is constant. In the last summand, only $\boldsymbol{\varepsilon}$ depends on the location, so the integration yields $V\bar{\boldsymbol{\varepsilon}}$. It remains only

$$F = \frac{1}{2V}\int_{\Omega} \bar{\boldsymbol{\varepsilon}}\mathbb{C}\boldsymbol{\varepsilon} \, \mathrm{d}V \tag{11.23}$$

$$= \frac{1}{2V} \int_{\Omega} \bar{\boldsymbol{\varepsilon}} \boldsymbol{\sigma} \, \mathrm{d}V \tag{11.24}$$

$$= \frac{1}{2} \bar{\boldsymbol{\varepsilon}} \bar{\boldsymbol{\sigma}} \leftarrow \bar{\boldsymbol{\sigma}} = \mathbb{C}^* \bar{\boldsymbol{\varepsilon}} \tag{11.25}$$

$$= \frac{1}{2} \bar{\boldsymbol{\varepsilon}} \mathbb{C}^* \bar{\boldsymbol{\varepsilon}}, \tag{11.26}$$

what corresponds to the effective strain energy density.

11.3 Maximum or minimum

The second variation of equation (11.7) is

$$\delta\delta F = \int_{\Omega} -\delta\boldsymbol{\tau} \Delta\mathbb{C}^{-1} \delta\boldsymbol{\tau} + \delta\tilde{\boldsymbol{\varepsilon}} \delta\boldsymbol{\tau} \, \mathrm{d}V. \tag{11.27}$$

We replace $\boldsymbol{\tau} = \Delta\boldsymbol{\sigma} - \mathbb{C}^0 \tilde{\boldsymbol{\varepsilon}}$ in the second term and find

$$\delta\delta F = \int_{\Omega} -\delta\boldsymbol{\tau} \Delta\mathbb{C}^{-1} \delta\boldsymbol{\tau} - \delta\tilde{\boldsymbol{\varepsilon}} \mathbb{C}^0 \delta\tilde{\boldsymbol{\varepsilon}} + \delta\tilde{\boldsymbol{\varepsilon}} \delta\Delta\boldsymbol{\sigma} \, \mathrm{d}V. \tag{11.28}$$

The last expression becomes zero with the same reasoning that has been applied between equations (11.15) and (11.17), such that

$$\delta\delta F = \int_{\Omega} -\delta\boldsymbol{\tau} \Delta\mathbb{C}^{-1} \delta\boldsymbol{\tau} - \delta\tilde{\boldsymbol{\varepsilon}} \mathbb{C}^0 \delta\tilde{\boldsymbol{\varepsilon}} \, \mathrm{d}V \tag{11.29}$$

remains.

11.3.1 Maximum condition

The second term is due to the positive definiteness of $\mathbb{C}^0$ smaller than zero. If $\Delta\mathbb{C}^{-1}$ is as well-positive definite everywhere, the first expression is also smaller than zero, which then ensures $\delta\delta F < 0$. This is the case if $\delta\mathbb{C}(\mathbf{x}) = \mathbb{C}(\mathbf{x}) - \mathbb{C}^0 > 0$ everywhere in the RVE, where the relational symbol ">" is meant in the sense of definiteness. Then F takes a maximum value at the solution.

11.3.2 Minimum condition

The proof of the minimum condition requires the following auxiliary integral, which is rewritten with $\delta\boldsymbol{\tau} = \delta\Delta\boldsymbol{\sigma} - \mathbb{C}^0 \delta\tilde{\boldsymbol{\varepsilon}}$,

$$\int_{\Omega} \delta\boldsymbol{\tau}\mathbb{C}^{0^{-1}}\delta\boldsymbol{\tau}\,\mathrm{d}V = \int_{\Omega} \delta\Delta\boldsymbol{\sigma}\mathbb{C}^{0^{-1}}\delta\Delta\boldsymbol{\sigma} - 2\delta\Delta\boldsymbol{\sigma}\delta\tilde{\boldsymbol{\varepsilon}} + \delta\tilde{\boldsymbol{\varepsilon}}\mathbb{C}^{0}\delta\tilde{\boldsymbol{\varepsilon}}\,\mathrm{d}V. \tag{11.30}$$

The expression $\delta\Delta\boldsymbol{\sigma}\delta\tilde{\boldsymbol{\varepsilon}}$ vanishes with the same reasoning as applied between equations (11.15) and (11.17), and it remains

$$\int_{\Omega} \delta\boldsymbol{\tau}\mathbb{C}^{0^{-1}}\delta\boldsymbol{\tau}\,\mathrm{d}V = \int_{\Omega} \delta\Delta\boldsymbol{\sigma}\mathbb{C}^{0^{-1}}\delta\Delta\boldsymbol{\sigma}\,\mathrm{d}V + \int_{\Omega} \delta\tilde{\boldsymbol{\varepsilon}}\mathbb{C}^{0}\delta\tilde{\boldsymbol{\varepsilon}}\,\mathrm{d}V. \tag{11.31}$$

Since $\mathbb{C}^0$ is positive definite, all summands are positive. Therefore,

$$\int_{\Omega} \delta\boldsymbol{\tau}\mathbb{C}^{0^{-1}}\delta\boldsymbol{\tau}\,\mathrm{d}V > \int_{\Omega} \delta\tilde{\boldsymbol{\varepsilon}}\mathbb{C}^{0}\delta\tilde{\boldsymbol{\varepsilon}}\,\mathrm{d}V. \tag{11.32}$$

If we replace in equation (11.29), the expression $\delta\tilde{\boldsymbol{\varepsilon}}\mathbb{C}^{0}\delta\tilde{\boldsymbol{\varepsilon}}$ by $\delta\boldsymbol{\tau}\mathbb{C}^{0^{-1}}\delta\boldsymbol{\tau}$, equation (11.29) becomes an inequality,

$$\delta\delta F > \int -\delta\boldsymbol{\tau}\Delta\mathbb{C}^{-1}\delta\boldsymbol{\tau} - \delta\boldsymbol{\tau}\mathbb{C}^{0^{-1}}\delta\boldsymbol{\tau}\,\mathrm{d}V. \tag{11.33}$$

We can now factor out $\delta\boldsymbol{\tau}$ and examine the fourth-order tensor,

$$\delta\delta F > \int \delta\boldsymbol{\tau}[-\Delta\mathbb{C}^{-1} - \mathbb{C}^{0^{-1}}]\delta\boldsymbol{\tau}\,\mathrm{d}V. \tag{11.34}$$

We consider the case $\mathbb{C}(\mathbf{x}) - \mathbb{C}^0 < 0$ everywhere in the sense of definiteness, complementary to the above maximum condition. We multiply the squared bracket from the left with $\mathbb{C}^0$ and from the right with $-\Delta\mathbb{C}$, which are both positive definite. This changes the value of $\delta\delta F$, but not the sign. This modification results in

$$\delta\delta F > \int \delta\boldsymbol{\tau}\underbrace{[\mathbb{C}^0 + \Delta\mathbb{C}]}_{\mathbb{C}}\delta\boldsymbol{\tau}\,\mathrm{d}V \tag{11.35}$$

Since $\mathbb{C}$ is positive definite everywhere, $\delta\delta F > 0$. Hence, F is a minimum for the choice $\mathbb{C}^0 > \mathbb{C}(\mathbf{x})$ everywhere.

11.4 Derivation of bounds

Since we know that depending on the choice of $\mathbb{C}^0$ the exact solution of the difference problem maximizes or minimizes F, we can use approximate solutions of the polarization stresses $\boldsymbol{\tau}$ to give values for F, which are strictly above or below the value at the solution, which we have identified as the effective strain energy. A sketch is given in Figure 11.1. Therefore, well-chosen approaches give bounds to the effective properties. For example, for isotropic phases with $K^{\pm}$ and $G^{\pm}$, one obtains with the larger values K^{+} and G^{+} in $\mathbb{C}^0$ an upper bound, and with the lower values K^{-} and G^{-} a lower bound.

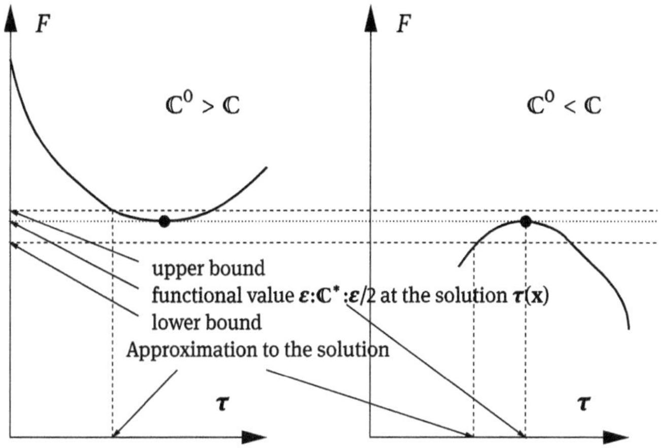

Figure 11.1: The choice of $\mathbb{C}^0$ controls the maximum or minimum condition. An approach for $\boldsymbol{\tau}$ gives then a bound to the solution.

11.5 Hashin–Shtrikman bounds

The Hashin–Shtrikman bounds are obtained by restricting the function space $\hat{\boldsymbol{\tau}}$ to some discrete set of parameters and minimizing or maximizing F over these parameters, very similar to the finite element method.

The approach that leads to the Hashin–Shtrikman bounds is to assume piecewise homogeneous polarization stresses $\boldsymbol{\tau}_i$ in regions with piecewise homogeneous stiffnesses $\mathbb{C}_i$,

$$\overline{\boldsymbol{\tau}} = \sum_{i=1,n} v_i \boldsymbol{\tau}_i. \tag{11.36}$$

The value of the functional is then

$$F = \frac{1}{2}\overline{\boldsymbol{\varepsilon}}\mathbb{C}^0\overline{\boldsymbol{\varepsilon}} - \sum_{i=1,n} \frac{v_i}{2}\boldsymbol{\tau}_i \Delta\mathbb{C}_i^{-1}\boldsymbol{\tau}_i + \frac{1}{2}\langle \boldsymbol{\tau}\tilde{\boldsymbol{\varepsilon}}\rangle + \overline{\boldsymbol{\tau}}\,\overline{\boldsymbol{\varepsilon}}. \tag{11.37}$$

Hashin and Shtrikman restricted their analysis in Hashin and Shtrikman (1963) to isotropic phases, which is why all second-order tensor fields can be decomposed into the dilatoric and deviatoric parts. Here, we stick to the notation with fourth-order tensors, and apply the dilator-deviator decomposition only at the end.[2] This allows applying the tensorial result to other cases than isotropic elasticity, like heat conduction in Section 11.6.

---

2 In Section 13.9.3, the Hashin–Shtrikman bounds are derived for phases with cubic stiffness and isotropic orientation distribution.

In $\langle \boldsymbol{\tau}\tilde{\boldsymbol{\varepsilon}}\rangle$ we can replace $\boldsymbol{\tau}$ by $\tilde{\boldsymbol{\tau}}$ without altering the value of the functional. We will see later as an exercise to Fourier methods in Section 12.5.1 that for isotropic microstructures and isotropic reference stiffnesses $\mathbb{C}^0$ we can write

$$\langle \boldsymbol{\tau}\tilde{\boldsymbol{\varepsilon}}\rangle = \langle \tilde{\boldsymbol{\tau}}\tilde{\boldsymbol{\varepsilon}}\rangle = -\langle \tilde{\boldsymbol{\tau}}\mathbb{W}\tilde{\boldsymbol{\tau}}\rangle \tag{11.38}$$

where the interaction tensor $\mathbb{W}$ depends only on the choice of $\mathbb{C}^0$. Using this, we obtain

$$F = \frac{1}{2}\bar{\boldsymbol{\varepsilon}}\mathbb{C}^0\bar{\boldsymbol{\varepsilon}} - \sum_{i=1,n}\frac{v_i}{2}\tilde{\boldsymbol{\tau}}_i\mathbb{W}\tilde{\boldsymbol{\tau}}_i - \sum_{i=1,n}\frac{v_i}{2}\boldsymbol{\tau}_i\Delta\mathbb{C}_i^{-1}\boldsymbol{\tau}_i + \bar{\boldsymbol{\tau}}\bar{\boldsymbol{\varepsilon}}. \tag{11.39}$$

We next mini- or maximize F over $\boldsymbol{\tau}_i$,

$$\mathbf{0} = \frac{\partial F}{\partial \boldsymbol{\tau}_j} = -\sum_{i=1,n} v_i\tilde{\boldsymbol{\tau}}_i\mathbb{W}\frac{\partial\tilde{\boldsymbol{\tau}}_i}{\partial\boldsymbol{\tau}_j} - \sum_{i=1,n} v_i\boldsymbol{\tau}_i\Delta\mathbb{C}_i^{-1}\frac{\partial\boldsymbol{\tau}_i}{\partial\boldsymbol{\tau}_j} + \frac{\partial\bar{\boldsymbol{\tau}}}{\partial\boldsymbol{\tau}_j}\bar{\boldsymbol{\varepsilon}}. \tag{11.40}$$

With $\bar{\boldsymbol{\tau}} = \sum v_i\boldsymbol{\tau}_i$, $\tilde{\boldsymbol{\tau}}_i = \boldsymbol{\tau}_i - \bar{\boldsymbol{\tau}}$ and $\partial\boldsymbol{\tau}_i/\partial\boldsymbol{\tau}_j = \delta_{ij}\mathbb{I}$ the result is

$$\mathbf{0} = -\sum_{i=1,n} v_i\tilde{\boldsymbol{\tau}}_i\mathbb{W}(\delta_{ij} - v_j) - \sum_{i=1,n} v_i\boldsymbol{\tau}_i\Delta\mathbb{C}_i^{-1}\delta_{ij} + v_j\bar{\boldsymbol{\varepsilon}} \tag{11.41}$$

$$= -v_j\tilde{\boldsymbol{\tau}}_j\mathbb{W} + v_j\mathbb{W}\underbrace{\sum_{i=1,n} v_i\tilde{\boldsymbol{\tau}}_i}_{\mathbf{0}} - v_j\boldsymbol{\tau}_j\Delta\mathbb{C}_j^{-1} + v_j\bar{\boldsymbol{\varepsilon}}. \tag{11.42}$$

$\mathbb{I}$ is the identity on symmetric second-order tensors. Since here all involved second-order tensors are symmetric, we can replace $\mathbb{I}$ by the scalar factor 1. Further, we can cancel out v_j in the last equation. It remains

$$(\mathbb{W} + \Delta\mathbb{C}_j^{-1})\tilde{\boldsymbol{\tau}}_j = \bar{\boldsymbol{\varepsilon}} \tag{11.43}$$

$$(\mathbb{W} + \Delta\mathbb{C}_j^{-1})\boldsymbol{\tau}_j = \bar{\boldsymbol{\varepsilon}} + \mathbb{W}\bar{\boldsymbol{\tau}}, \tag{11.44}$$

which is a linear system for $\boldsymbol{\tau}_j$. The coupling takes place over $\bar{\boldsymbol{\tau}}$ in $\tilde{\boldsymbol{\tau}}_j = \boldsymbol{\tau}_j - \bar{\boldsymbol{\tau}}$. We can solve formally for $\boldsymbol{\tau}_j$ and insert this in $\bar{\boldsymbol{\tau}}$, which gives simple expressions for the $\boldsymbol{\tau}_j$,

$$\boldsymbol{\tau}_j = \underbrace{(\mathbb{W} + \Delta\mathbb{C}_j^{-1})^{-1}}_{\mathbb{B}_j}(\bar{\boldsymbol{\varepsilon}} + \mathbb{W}\bar{\boldsymbol{\tau}}). \tag{11.45}$$

This can be used to determine $\bar{\boldsymbol{\tau}}$,

$$\bar{\boldsymbol{\tau}} = \sum v_i\boldsymbol{\tau}_i = \underbrace{\sum(v_i\mathbb{B}_i)}_{\mathbb{A}}(\bar{\boldsymbol{\varepsilon}} + \mathbb{W}\bar{\boldsymbol{\tau}}), \tag{11.46}$$

which is used to write $\bar{\boldsymbol{\tau}}$ as a function of $\bar{\boldsymbol{\varepsilon}}$,

$$(\mathbb{I} - \mathbb{A}\mathbb{W})\bar{\boldsymbol{\tau}} = \mathbb{A}\bar{\boldsymbol{\varepsilon}}, \tag{11.47}$$

$$\bar{\boldsymbol{\tau}} = \underbrace{(\mathbb{I} - \mathbb{A}\mathbb{W})^{-1}\mathbb{A}}_{\mathbb{L}}\,\bar{\boldsymbol{\varepsilon}}. \tag{11.48}$$

This is finally used to obtain the solution for $\boldsymbol{\tau}_j$,

$$\boldsymbol{\tau}_j = \underbrace{\mathbb{B}_j(\mathbb{I} + \mathrm{W}\mathbb{L})}_{\mathbb{L}_j}\bar{\boldsymbol{\varepsilon}}. \tag{11.49}$$

Here, $\mathbb{L}$ is not the stress concentration tensor but just an intermediate quantity. We use this to replace in the functional all $\boldsymbol{\tau}$-expressions by $\bar{\boldsymbol{\varepsilon}}$ and identify the effective stiffness,

$$F = \frac{1}{2}\bar{\boldsymbol{\varepsilon}}\underbrace{\left[\mathbb{C}^0 - \sum_{i=1,n} v_i(\mathbb{L}_i - \mathbb{L})\mathrm{W}(\mathbb{L}_i - \mathbb{L}) - \sum_{i=1,n} v_i\mathbb{L}_i\Delta\mathbb{C}_i^{-1}\mathbb{L}_i + 2\mathbb{L}\right]}_{\mathbb{C}_{\mathrm{HS}}}\bar{\boldsymbol{\varepsilon}}, \tag{11.50}$$

which can be summarized with $\sum v_i\mathbb{L}_i = \mathbb{L}$ to

$$\mathbb{C}_{\mathrm{HS}} = \mathbb{C}^0 - \sum_{i=1,n} v_i\mathbb{L}_i(\mathrm{W} + \Delta\mathbb{C}_i^{-1})\mathbb{L}_i + \mathbb{L}\mathrm{W}\mathbb{L} + 2\mathbb{L}. \tag{11.51}$$

This can be implemented relatively easily, representing fourth-order tensors with subsymmetries as 6×6 matrices. Moreover, all tensors are isotropic and, therefore, coaxial, i. e., the factors are commutative, and all can be reduced to scalar coefficients that belong to the isotropic projectors. An example implementation is given in Listing 11.1.

Listing 11.1: Mathematica notebook (https://gitlab.com/gluegerainer/listings-homogenization-methods/-/blob/main/Listing-13_en.nb) to determine the Hashin–Shtrikman bounds.

```
Remove["Global`*"] (* Revoke all declarations *)
Translate3333to66[C3333_] := ( (* Aux. function to convert W_ijkl to a 6x6 matrix *)
  Ivek1 = {1, 2, 3, 1, 1, 2};
  Ivek2 = {1, 2, 3, 2, 3, 3};
  Faktvek = {1, 1, 1, Sqrt[2], Sqrt[2], Sqrt[2]};
  Table[C3333[[Ivek1[[i]], Ivek2[[i]], Ivek2[[j]], Ivek1[[j]]]] Faktvek[[i]]*Faktvek[[j]], {i,
      1, 6}, {j, 1, 6}])
id = IdentityMatrix[3]; (* second-order identity *)
I6 = IdentityMatrix[6]; (* fourth-order identity *)

(* Set up interaction tensor W *)
W1 = 1/15 Table[id[[i, j]] id[[k, l]] + id[[i, l]] id[[k, j]] + id[[i, k]] id[[j, l]], {i, 3}, {
      j, 3}, {k, 3}, {l, 3}];
W2base = 1/3 Table[id[[i, l]] id[[k, j]], {i, 3}, {j, 3}, {k, 3}, {l, 3}];
W2aux1 = 1/2 (W2base + Transpose[W2base, 3 <-> 4]); (* Symmetrization *)
W2 = 1/2 (W2aux1 + Transpose[W2aux1, 1 <-> 2]);
a = (1/(K0 + 4/3 G0) - 1/G0);
W = Translate3333to66[a W1 + 1/G0 W2] // FullSimplify; (* Store as 6x6 matrix *)

(* Projectors, stiffnesses, and inverted stiffness-differences *)
P1 = TensorProduct[{1, 1, 1, 0, 0, 0}, {1, 1, 1, 0, 0, 0}]/3;
P2 = I6 - P1;
C0 = 3 K0 P1 + 2 G0 P2;
C1 = 3 K1 P1 + 2 G1 P2;
C2 = 3 K2 P1 + 2 G2 P2;
iDC1 = Inverse[C1 - C0] // FullSimplify;
```

```
iDC2 = Inverse[C2 - C0] // FullSimplify;

(* Use abbreviations *)
B1 = Inverse[iDC1 + W] // FullSimplify;
B2 = Inverse[iDC2 + W] // FullSimplify;
A = ((1 - v2) B1 + v2 B2) // FullSimplify;
L = Inverse[Inverse[A] - W] // Simplify; (* Tau_bar = L  Eps_bar*)
L1 = B1.(I6 + W.L) // Simplify;(* Tau_1 = L1  Eps_bar*)
L2 = B2.(I6 + W.L) // Simplify; (* Tau_2 = L2  Eps_bar*)

(* HS-stiffness *)
CHS = C0 - (1 - v2) Transpose[L1].(W + iDC1).L1 - v2 Transpose[L2].(W + iDC2).L2 + Transpose[L].
      W.L + 2 L;

(* Extract G and K *)
GHS = CHS[[6, 6]]/2 // Simplify;
KHS = CHS[[1, 1]] - 4/3 GHS // Simplify

(* G and K according to H and S as functions of G0 and K0 *)
(* The numbers are taken from  *)
(* Z. Hashin, S. Shtrikman, A VARIATIONAL APPROACH TO THE THEORY OF ELASTIC BEHAVIOUR *)
(* OF MULTIPHASE MATERIALS, J. Mech. Phys. Solids 1963 (11) 127-140 *)
K1 = 25.0;
K2 = 60.7;
G1 = 11.5;
G2 = 41.8;
KHSf[v2_, K0_, G0_] = KHS // Simplify // Chop;
GHSf[v2_, K0_, G0_] = GHS // Simplify // Chop;

(* Plot results *)
v1 = 1 - v2;
G1HS = GHSf[v2, K1, G1] // Simplify
G2HS = GHSf[v2, K2, G2] // Simplify
HSpaperG1 = G1 + v2/(1/(G2 - G1) + 6 (K1 + 2 G1) v1/5/G1/(3 K1 + 4 G1)) // Simplify
HSpaperG2 = G2 + v1/(1/(G1 - G2) + 6 (K2 + 2 G2) v2/5/G2/(3 K2 + 4 G2)) // Simplify
Plot[{G1HS, G2HS, HSpaperG1, HSpaperG2, v1 G1 + v2 G2, 1/(v1/G1 + v2/G2)}, {v2, 0, 1}]
K1HS = KHSf[v2, K1, G1] // Simplify
K2HS = KHSf[v2, K2, G2] // Simplify
HSpaperK1 = K1 + v2/(1/(K2 - K1) + 3 v1/(3 K1 + 4 G1)) // Simplify
HSpaperK2 = K2 + v1/(1/(K1 - K2) + 3 v2/(3 K2 + 4 G2)) // Simplify
Plot[{K1HS, K2HS, HSpaperK1, HSpaperK2, v1 K1 + v2 K2, 1/(v1/K1 + v2/K2)}, {v2, 0, 1}]
```

The listing produces the graphs of the Hashin–Shtrikman bounds in comparison to the Voigt–Reuss bounds; see Figure 11.2. To finally obtain simple expressions for K_{HS} and G_{HS}, we use the projector representation, i. e., the dilator-deviator decomposition. Scalar multiplication of equation (11.51) with the two isotropic projectors gives two scalar equations for K_{HS} and G_{HS} that contain some nested sums and inversions. Simplifying these, one finds

$$3K_{\mathrm{HS}} = 3K_0 + \left(\kappa_K^{-1} - w_1\right)^{-1}, \quad \kappa_K = \sum \frac{v_i}{(3K_i - 3K_0)^{-1} + w_1}, \quad w_1 = \frac{1}{3K_0 + 4G_0}, \tag{11.52}$$

$$2G_{\mathrm{HS}} = 2G_0 + \left(\kappa_G^{-1} - w_2\right)^{-1}, \quad \kappa_G = \sum \frac{v_i}{(2G_i - 2G_0)^{-1} + w_2}, \quad w_2 = \frac{3(K_0 + 2G_0)}{5(3K_0 G_0 + 4G_0^2)}. \tag{11.53}$$

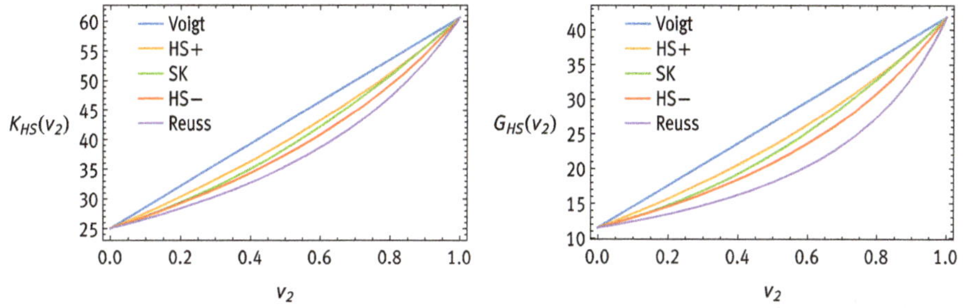

Figure 11.2: The Hashin–Shtrikman bounds in comparison to the Voigt–Reuss bounds and the self-consistent estimate. The numerical values in Hashin and Shtrikman (1963) for tungsten-carbide (K_1 = 25.0 GPa, K_2 = 60.7 GPa, G_1 = 11.5 GPa, G_2 = 41.8 GPa) have been used.

The scalars w_1 and w_2 are the eigenvalues of the interaction tensor $\mathbb{W} = w_1\mathbb{P}_{I1} + w_2\mathbb{P}_{I2}$ (equation (12.57)).

11.6 The Hashin–Shtrikman bounds for heat conduction and diffusion

The above derivation can be specified for isotopic linear constitutive laws of the form

$$\mathbf{q} = \mathbf{L} \cdot \mathbf{g}, \quad \mathbf{L} = l\mathbf{I} \tag{11.54}$$

for vector quantities like heat flux and temperature gradient or diffusion vector and concentration gradient. There are two interesting cases: an isotropic material distribution in the 3D space ($\mathbb{R}_3$) and in a 2D plane ($\mathbb{R}_2$). The latter can be viewed as transverse isotropy when an embedding in the 3D space is considered, with the axis of transverse isotropy being the plane's normal. Ignoring this embedding, we can speak of isotropy in $\mathbb{R}_2$.

To determine the Hashin–Shtrikman bounds, we merely need to insert the volume fractions v_i, the constitutive tensors $\mathbf{L}_{i\,2,3} = l_i\mathbf{I}_{2,3}$ as well as the second-order interaction tensors $\mathbf{W}_{2,3}$ into equation (11.51). These are

$$\mathbf{W}_D = \frac{1}{Dl_0}\mathbf{I} \tag{11.55}$$

with the reference conductivity l_0 and the dimension D of the vector space $D = 2$ or $D = 3$. $\mathbf{W}_{2,3}$ are derived in Section 12.5.2. Inserting all of this into equation (11.51), one sees that all involved tensors are multiples of the identity tensor, which allows switching to scalar equations immediately. The result is

$$l_{HS} = l_0 - (1 + w\underline{l})^2 a + \underline{l}^2 w + 2\underline{l} \tag{11.56}$$

$$\underline{l} = (a^{-1} - w)^{-1} \tag{11.57}$$

$$a = \sum v_i b_i \tag{11.58}$$

$$b_i = \left(w + (l_i - l_0)^{-1}\right)^{-1} \tag{11.59}$$

$$w = (Dl_0)^{-1}. \tag{11.60}$$

Evaluated for two conductivities l_1 and l_2, we obtain

$$l_{HS} = \frac{l_1 l_2 + (D-1) l_0 (l_2 v_2 - l_1 v_1)}{(D-1) l_0 + l_1 v_2 + l_2 v_1}. \tag{11.61}$$

The bounds are obtained by setting the reference conductivity l_0 to the largest or smallest conductivity $l_0 = \min(l_1, l_2 \ldots l_n)$ and $l_0 = \max(l_1, l_2 \ldots l_n)$. Comparisons of the tighter Hashin–Shtrikman bounds to the Voigt–Reuss bounds are given in Figure 11.3 for $D = 2$ and $D = 3$. One sees that the Hashin–Shtrikman bounds are a noteworthy improvement of the Voigt–Reuss bounds. The HS bounds are for $D = 3$ closer to the upper Voigt bound, and the HS bounds for $D = 2$ are closer to the lower Reuss bound. In fact, setting $D = 1$ in equation (11.61) one obtains the lower Reuss bound, and evaluating the limit $D \to \infty$ of equation (11.61), one obtains the upper Voigt bound. This is in accordance with the discussion in Section 7.1.7, where we noted that the Reuss bound gives the true effective value for $D = 1$, and the Voigt bound gives the true effective value for $D \to \infty$.

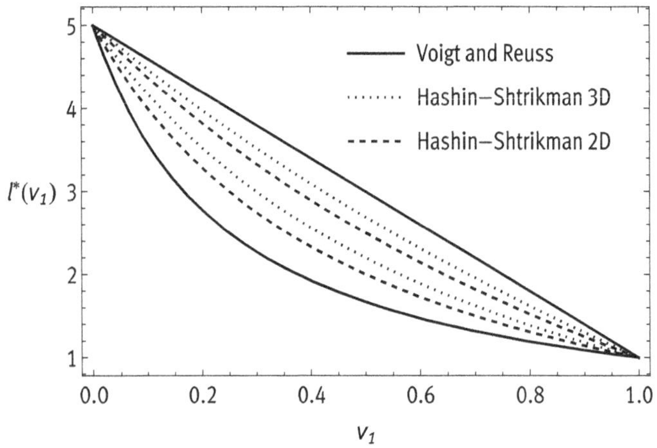

Figure 11.3: Effective conductivities l^* over the volume fraction v_1 according to the Hashin–Shtrikman bounds for vectorial constitutive laws in case of planar and spatial isotropy in comparison to the Voigt–Reuss bounds with $l_1 = 1$ and $l_2 = 5$.

Listing 11.2: Mathematica notebook (https://gitlab.com/gluegerainer/listings-homogenization-methods/-/blob/main/Listing-14_en.nb) for plotting the Hashin–Shtrikman bounds for isotropic vector problems in Figure 11.3.

```
Remove["Global`*"]
W = 1/L0/dim;
v2 = 1 - v1;
Li = {L1, L2};
Bi = 1/(W + 1/(Li - L0));
A = v1 Bi[[1]] + v2 Bi[[2]];
L = 1/(1/A - W);
Lhs = L0 - (1 + W L)^2 A + L^2 W + 2 L;
plotfunctions = {L1 v1 + L2 v2, Lhs /. {L0 -> L2, dim -> 3},
    Lhs /. {L0 -> L2, dim -> 2}, Lhs /. {L0 -> L1, dim -> 3},
    Lhs /. {L0 -> L1, dim -> 2}, 1/(v1/L1 + v2/L2)} /. {L1 -> 1,
    L2 -> 5};
Plot[plotfunctions, {v1, 0, 1},
 PlotLegends -> {"Voigt", "HS up 3D", "HS up 2D", "HS low 3D", "HS low 2D", "Reuss"}]
```

11.7 Comments regarding the Hashin–Shtrikman bounds

One can see in Figures 11.2 and 11.3 that the effort to derive the Hashin–Shtrikman bounds paid off. They are much tighter than the Voigt–Reuss bounds. They are, in fact, the tightest possible bounds that can be obtained for isotropic microstructures. The bounds are attained by special microstructures, namely the coated sphere model depicted in Figure 11.4. This microstructure consists of spherical inclusions with a surrounding spherical shell with a fixed ratio of radii, which fill out the entire volume.

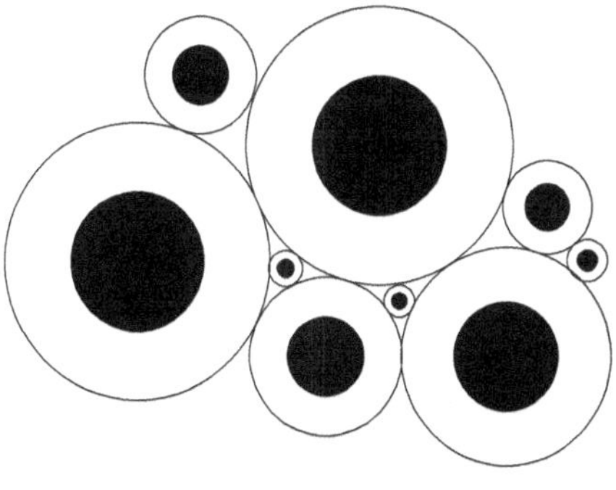

Figure 11.4: The coated sphere model.

If the Hashin–Shtrikman bounds are violated experimentally, this just means that one of the underlying assumptions is not satisfied: The microstructure or the phases are not isotropic. This is why the Voigt–Reuss bounds are still useful: They also hold for anisotropic arrangements of isotropic phases.

All estimates that have been presented in Sections 9.4 to 9.6 (Mori–Tanaka, differential scheme, self-consistent method) lie between the Hashin–Shtrikman bounds (differential scheme, self-consistent method) or on the lower Hashin–Shtrikman bound (Mori–Tanaka). The self-consistent method approaches at extreme volume fractions the Hashin–Shtrikman bound that has the smaller absolute slope (Figure 11.2). This reflects the assumed minimal interaction among inclusions at small volume fractions, which is baked into the Eshelby solution while treating both phases equally. One can use other basic solutions to obtain different estimates. For example, for approaches with the converse behavior, i. e., that approach the Hashin–Shtrikman bound with the larger slope at extreme volume fractions, one can take the orientation average of the basic solution for laminates (Kalisch and Glüge, 2015). Hence, when choosing an approach, one should always check whether the underlying assumptions suit at least approximately the microstructure in question.

12 Fourier and Green methods

The methods presented so far rely on one or more of the following: ignoring the microstructure, special basic solutions for idealized microstructures or solving specific boundary value problems. We now consider the representation of fields in Fourier[1] space. This representation extends our toolbox greatly because

- The periodic repeatability of a virtual material sample is provided automatically in Fourier space since the function basis is periodic.
- An isotropy of the microstructure becomes a symmetry in Fourier space (Section 12.4). This allows, e. g., to determine the interaction tensor that appears in the derivation of the Hashin–Shtrikman bounds (Section 12.5.1) or the fundamental solution to isotropic right-hand side perturbation functions (Section 12.8).
- In Fourier space, fields can be decomposed algebraically into the homogeneous part, the divergence-free part and the rotation-free part. This can be used to give closed-form expressions for the effective material properties (Section 12.6).
- The fast discrete Fourier transform allows for an efficient numerical solution of periodic boundary value problems without setting up a large linear system. This method is memory efficient compared to standard methods such as FEM and FDM, hence much finer discretizations are possible (Section 12.7).

12.1 Fourier series of real functions

Let

$$[f(x), g(x)] = \frac{1}{2\pi} \int_{-\pi}^{\pi} f(x) g^{\dagger}(x)\,\mathrm{d}x \tag{12.1}$$

be the definition of the scalar product between two functions $f(x)$ and $g(x)$, where $g^{\dagger}(x)$ is the complex conjugate of $g(x)$. We now consider all functions

$$f_k(x) = \mathrm{e}^{\mathrm{i}kx} \quad \text{with } k \in \mathbb{Z}. \tag{12.2}$$

With $f_k^{\dagger}(x) = \mathrm{e}^{-\mathrm{i}kx}$ and the conversion

$$\mathrm{e}^{\mathrm{i}\phi} = \cos\phi + \mathrm{i}\sin\phi, \tag{12.3}$$

it is easy to see that

$$[f_k(x), f_l(x)] = \frac{1}{2\pi} \int_{-\pi}^{\pi} \mathrm{e}^{\mathrm{i}kx} \mathrm{e}^{-\mathrm{i}lx}\,\mathrm{d}x \tag{12.4}$$

1 Jean Baptiste Joseph Fourier, 1768–1830 (https://en.wikipedia.org/wiki/Joseph_Fourier).

https://doi.org/10.1515/9783110793529-012

$$= \frac{1}{2\pi} \int_{-\pi}^{\pi} \mathrm{e}^{\mathrm{i}(k-l)x}\, \mathrm{d}x \tag{12.5}$$

$$= \frac{1}{2\pi} \int_{-\pi}^{\pi} \cos((k-l)x) + \mathrm{i}\sin((k-l)x)\, \mathrm{d}x \tag{12.6}$$

$$= \delta_{kl}, \tag{12.7}$$

since for $k \neq l$, $k, l \in \mathbb{Z}$, always entire periods are integrated. Only for $k = l$, the integration of $\cos 0 = 1$ gives 2π, the sine part $\sin 0$ also vanishes for this case. Therefore, $f_k(x) = \mathrm{e}^{\mathrm{i}kx}$ is an orthonormal basis for functions defined between $-\pi$ and π, which allows for the representation[2]

$$f(x) = \sum_{k=-\infty,\infty} \mathrm{e}^{\mathrm{i}kx} \hat{f}_k. \tag{12.8}$$

The components w. r. t. this basis are obtained with the above-defined scalar product

$$\hat{f}_k = [f(x), f_k^\dagger(x)]. \tag{12.9}$$

As an example, consider the sawtooth wave $f(x) = \mathrm{mod}(x + \pi, 2\pi) - \pi$. For integrating between $-\pi$ and π, it is sufficient to use $f(x) = x$. The components $\hat{f}_k$ w. r. t. the Fourier basis is

$$\hat{f}_0 = \frac{1}{2\pi} \int_{-\pi}^{\pi} x\, \mathrm{d}x = 0, \tag{12.10}$$

$$\hat{f}_1 = \frac{1}{2\pi} \int_{-\pi}^{\pi} x\mathrm{e}^{-\mathrm{i}x}\, \mathrm{d}x = -\mathrm{i}, \tag{12.11}$$

$$\hat{f}_{-1} = \frac{1}{2\pi} \int_{-\pi}^{\pi} x\mathrm{e}^{\mathrm{i}x}\, \mathrm{d}x = \mathrm{i}, \tag{12.12}$$

$$\hat{f}_2 = \frac{1}{2\pi} \int_{-\pi}^{\pi} x\mathrm{e}^{-2\mathrm{i}x}\, \mathrm{d}x = \mathrm{i}/2, \tag{12.13}$$

$$\hat{f}_{-2} = \frac{1}{2\pi} \int_{-\pi}^{\pi} x\mathrm{e}^{2\mathrm{i}x}\, \mathrm{d}x = -\mathrm{i}/2, \tag{12.14}$$

$$\vdots$$

We use these components to approximate the sawtooth wave,

2 We ignore a lot of technical details here.

$$f(x) \approx \hat{f}_0 + \hat{f}_1 e^{ix} + \hat{f}_{-1} e^{-ix} + \hat{f}_2 e^{i2x} + \hat{f}_{-2} e^{-2ix} \dots \tag{12.15}$$

The approximation improves with an increasing number of Fourier components. In Figure 12.1, the approximation is drawn up to $k = \pm 3$. The Fourier coefficients have no real part due to the sawtooth wave being an odd function with $f(-x) = -f(x)$. The real part corresponds to the cosine term in equation (12.3), for which $\cos(x) = \cos(-x)$ holds. The cosines construct the even part of the function that is approximated. The imaginary part corresponds to the sine term in equation (12.3), for which $\sin(-x) = -\sin(x)$ holds. These construct the odd part of $f(x)$.

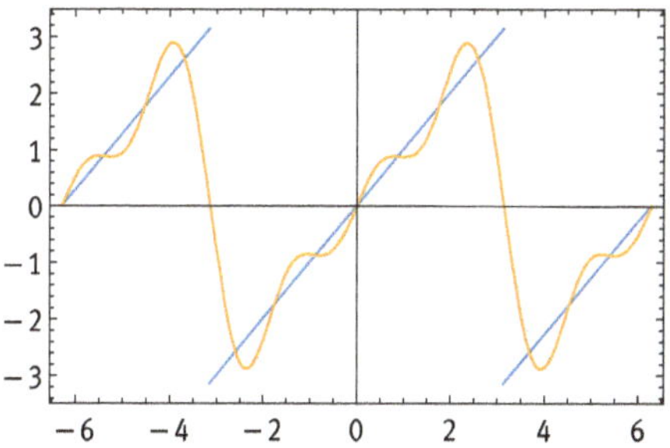

Figure 12.1: Approximation of a sawtooth wave with a Fourier series up to the third order. Dropping the higher, less perceivable frequencies is used in lossy compression formats.

12.2 Fourier series in $\mathbb{R}_3$

The generalization of the Fourier series from real functions to functions defined on $\mathbb{R}_3$ is achieved by extending the basis function to the plane wave approach

$$f_{\mathbf{k}}(\mathbf{x}) = e^{i\mathbf{k}\cdot\mathbf{x}} \quad \text{with } k_i \in \mathbb{Z}. \tag{12.16}$$

$\mathbf{k}$ is the wave vector, which has only integer components. $\mathbf{k}$ gives the direction of a plane wave. The magnitude $|\mathbf{k}|$ is the reciprocal of the wavelength divided by 2π, hence the frequency. The scalar product becomes

$$[f_{\mathbf{k}}(\mathbf{x}), f_{\mathbf{g}}(\mathbf{x})] = \frac{1}{(2\pi)^3} \int_{-\pi}^{\pi} \int_{-\pi}^{\pi} \int_{-\pi}^{\pi} f_{\mathbf{k}}(\mathbf{x}) f_{\mathbf{g}}^{\dagger}(\mathbf{x})\, dx_3\, dx_2\, dx_1. \tag{12.17}$$

As in the previous section, one can easily check that

$$[f_{\mathbf{k}}(\mathbf{x}), f_{\mathbf{g}}(\mathbf{x})] = \delta_{\mathbf{kg}} \tag{12.18}$$

with

$$\delta_{\mathbf{kg}} = \begin{cases} k_1 = g_1, k_2 = g_2, k_3 = g_3 : 1 \\ \text{otherwise} : 0. \end{cases} \tag{12.19}$$

An approximation of a scalar function defined on $\mathbb{R}_3$ (a field) by a Fourier series is then

$$f(\mathbf{x}) = \sum_{k_1,k_2,k_3=-\infty\cdots\infty} e^{i\mathbf{k}\cdot\mathbf{x}} \hat{f}_{\mathbf{k}} \tag{12.20}$$

$$= \sum_{k_1,k_2,k_3=-\infty\cdots\infty} e^{ik_1x_1} e^{ik_2x_2} e^{ik_3x_3} \hat{f}_{\mathbf{k}} \tag{12.21}$$

$$= \sum_{k_1=-\infty\cdots\infty} e^{ik_1x_1} \sum_{k_2=-\infty\cdots\infty} e^{ik_2x_2} \sum_{k_3=-\infty\cdots\infty} e^{ik_3x_3} \hat{f}_{\mathbf{k}}. \tag{12.22}$$

The Fourier coefficients are calculated as before with the scalar product

$$\hat{f}_{\mathbf{k}} = [f(\mathbf{x}), f_{\mathbf{k}}(\mathbf{x})]. \tag{12.23}$$

Like in the case of real functions, this corresponds to a superposition of sine and cosine waves, only that in dimensions larger than one, a plane wave direction needs to be added via **k**. An illustrative video of this superposition (beginning after minute 42) can be found at https://www.youtube.com/watch?v=a7TUIkn3qjY&t=42m.

12.3 From Fourier series to Fourier transforms

A Fourier series of length n assigns n Fourier coefficients to a real function. These can be identified uniquely with n function values.

One obtains the Fourier transform when going from discrete integer frequencies to real frequencies,

$$\hat{f}(k) = \frac{1}{2\pi} \int_{\mathbb{R}} f(x) e^{ikx} \, dx. \tag{12.24}$$

Instead of integer indices k, the real value $k \in \mathbb{R}$ is the argument of the transformed function. Moreover, the integration domain is infinitely large, including $k \to \infty$, i. e., very short wavelengths and high frequencies. The frequency $k = 0$ corresponds to the constant, homogeneous part (i. e., an infinite wavelength), and all nonzero frequencies belong to sine or cosine waves that oscillate around zero. *Therefore, we have the transition from periodic, finite domains to aperiodic, infinite domains when going from discrete Fourier series to continuous Fourier transforms.* The inverse Fourier transform is

$$f(x) = \int_{\mathbb{R}} \hat{f}(k) e^{-ikx} \, dk. \tag{12.25}$$

It is not easy to see that these operations are inverse to each other. Inserting, for example, $f(x) = 1$, the forward and backward transformations are

$$f(x) = \frac{1}{2\pi} \int_{\mathbb{R}} \frac{e^{-ikx} e^{ikx}}{ik} \, dk = \frac{1}{2\pi} \int_{\mathbb{R}} \frac{1}{ik} \, dk, \tag{12.26}$$

which is a divergent integral due to the pole at $k = 0$. For its evaluation, one can use contour integration in the complex plane or the residue theorem. Contour integration of analytical functions is achieved by converting the integral along the real axis into a ring integral in the complex plane along the real axis and a closing arc at infinity. For analytical functions, this ring can be contracted without changing its value, as long as no poles are crossed. The result is zero if no poles are contained and nonzero for individual poles. A fundamental result is

$$\oint \frac{1}{k} \, dk = \pm 2\pi i, \tag{12.27}$$

where integration is done along a closed contour around the pole at $k = 0$. The sign depends on the direction of circulation. With this, one obtains again $f(x) = 1$ in equation (12.26).

This explains the origin of the factor $\frac{1}{2\pi}$ even though the domain is infinitely large instead of the interval $-\pi$ to π. The positive sign results from the counterclockwise direction of circulation. In d dimensions, this contour integration needs to be carried out d times, which is why the Fourier transform, and its inverse in d dimensions are

$$\hat{f}(\mathbf{k}) = \frac{1}{(2\pi)^{d/2}} \int_{\mathbb{R}_d} f(\mathbf{x}) e^{i\mathbf{k}\cdot\mathbf{x}} \, dv, \tag{12.28}$$

$$f(\mathbf{x}) = \frac{1}{(2\pi)^{d/2}} \int_{\mathbb{R}_d} \hat{f}(\mathbf{k}) e^{-i\mathbf{k}\cdot\mathbf{x}} \, dv_{\mathbf{k}}. \tag{12.29}$$

We have distributed the factor $\frac{1}{(2\pi)^d}$ evenly on forward and backward transform, but other conventions for the normalizing factor are possible. The even distribution on both transforms has the advantage that no prefactor appears in the Plancherel[3]–Parseval[4]–Rayleigh[5] theorem, which helps us to convert integrals over $\mathbb{R}_3$ in real space into integrals over $\mathbb{R}_3$ in Fourier space,

$$\int_{\mathbb{R}_3} \hat{f}(\mathbf{k}) \hat{g}^{\dagger}(\mathbf{k}) \, dv_{\mathbf{k}} = \int_{\mathbb{R}_3} f(\mathbf{x}) g^{\dagger}(\mathbf{x}) \, dv. \tag{12.30}$$

3 Michel Plancherel, 1885–1967 (https://en.wikipedia.org/wiki/Michel_Plancherel).

4 Marc-Antoine Parseval, 1755–1836 (https://en.wikipedia.org/wiki/Marc-Antoine_Parseval).

5 John Strutt, 3. Baron Rayleigh, 1842–1919 (https://en.wikipedia.org/wiki/John_William_Strutt,_3rd_Baron_Rayleigh).

Likewise, the change of sign can be assigned to the forward or backward transform. Here, we choose the variant above.

We will use both Fourier series and Fourier transforms:

- We examine specific boundary value problems of periodically repeatable RVE with microstructure information at Cartesian grid points. Then the representation in Fourier space is discrete and periodic.
- For aperiodic, infinitely large RVE, we use the Fourier transformation to derive expressions for the effective properties. Then the Fourier representation is also infinitely large and aperiodic, and we make use of its Fourier space properties like symmetries and decompositions that are not available in real space.

12.4 Properties of the Fourier coefficients

12.4.1 Homogeneous part

It is easy to see that the Fourier coefficient to $k = 0$ in $\mathbb{R}_1$ and $\mathbf{k} = \{0, 0, 0\}$ in $\mathbb{R}_3$ corresponds to the constant or homogeneous part of the field, which is equal to the mean value because all other Fourier coefficients are the amplitudes of (co)sine waves that oscillate around this value. The latter are called the fluctuating part of the field.

12.4.2 Symmetries in Fourier space reflect statistical properties of the microstructure

Statistical properties of the microstructure in real space manifest as symmetries in Fourier space. For example, the Fourier space representation of a statistically isotropic field in real space has spherical symmetry in Fourier space.

Moreover, the Fourier coefficients with small wave numbers k or $|\mathbf{k}|$ belong to long periodic regularities. Large wave numbers represent short periodic regularities of a field. This is depicted in Figure 12.2. In the first row of images, an indicator function for small circular inclusions that are homogeneously distributed is depicted, where the field is zero in the black inclusions and the field value is 1 in the white matrix phase. The corresponding Fourier coefficients show an almost perfect radial symmetry around the coefficient to $\mathbf{k} = \{0, 0\}$. The latter is the largest because the white phase (value 1) dominates the microstructure. In the second row of images, we have homogeneously distributed small squares that are aligned parallel to each other. Therefore, the Fourier coefficients that belong to small wave numbers (low frequencies, large wavelengths) in the right figure appear almost isotropic, while for higher frequencies (high frequencies, short wavelengths), the anisotropy of the aligned squares becomes apparent. The microstructure in the bottom row consists of circular inclusions, which are distributed unevenly in the horizontal direction. Therefore, the spectrum appears

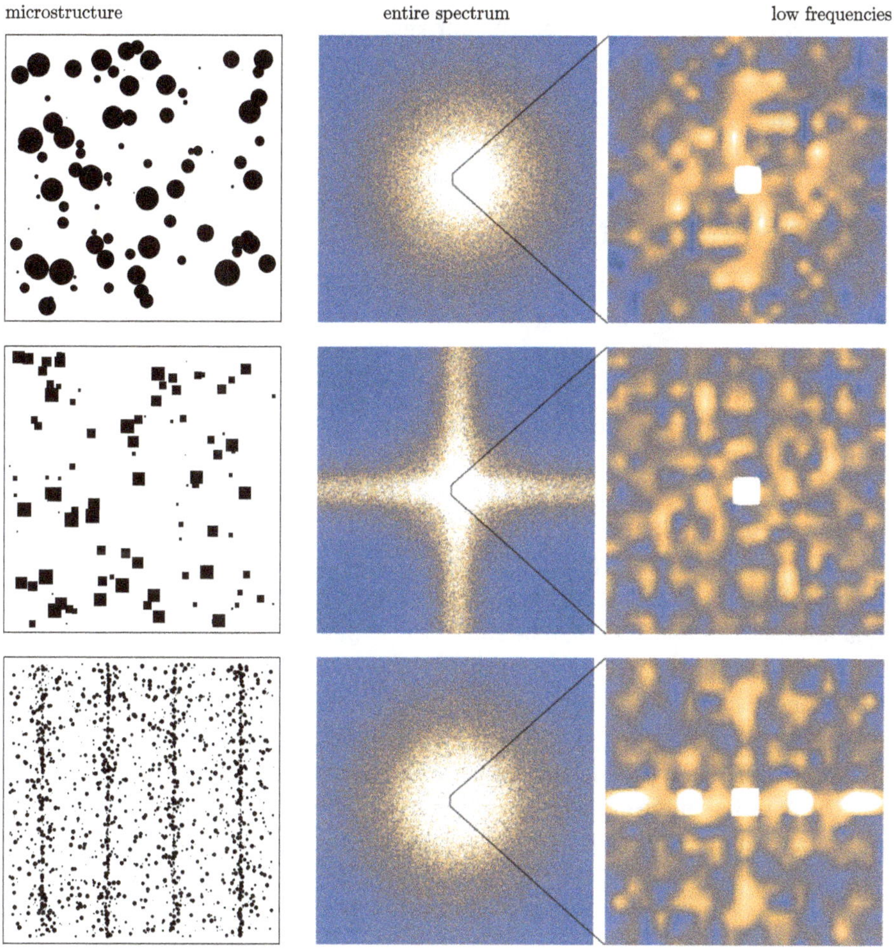

Figure 12.2: Microstructures (left), their entire Fourier spectrum (center) and a magnification of the low frequency/long wavelength Fourier coefficients (right), for which the color scale has been adapted.

almost isotropic for the high frequencies (middle image), but a strong anisotropy is visible in the lower frequencies (right image).

12.5 Hashin–Shtrikman interaction tensor

12.5.1 Linear elasticity

We have seen in the last section that statistically isotropic microstructures have a Fourier spectrum with spherical symmetry in 3D or radial symmetry in 2D. This property is used when deriving the Hashin–Shtrikman bounds. When evaluating the HS-functional, the expression

$$\int_{\Omega} \boldsymbol{\tau}(\mathbf{x}) : \tilde{\boldsymbol{\varepsilon}}(\mathbf{x})\,\mathrm{d}V = \int_{\Omega} \tilde{\boldsymbol{\tau}} : \tilde{\boldsymbol{\varepsilon}}\,\mathrm{d}V \tag{12.31}$$

appears. As the solution, we have

$$(\mathbb{C}^0 : \tilde{\boldsymbol{\varepsilon}}(\mathbf{x}) + \boldsymbol{\tau}(\mathbf{x})) \cdot \nabla = \mathbf{o} \quad \text{with } \tilde{\boldsymbol{\varepsilon}}(\mathbf{x}) = \tilde{\mathbf{u}}(\mathbf{x}) \otimes \nabla. \tag{12.32}$$

The symmetrization of $\tilde{\mathbf{u}}(\mathbf{x}) \otimes \nabla$ is contained in $\mathbb{C}^0$. We consider an infinitely large material sample, i. e., the entire $\mathbb{R}_3$ instead of a periodic RVE. With Plancherel's theorem, we can write the integral over the real space as an integral over the Fourier space,

$$\int_{\mathbb{R}_3} \boldsymbol{\tau}(\mathbf{x}) : \tilde{\boldsymbol{\varepsilon}}(\mathbf{x})\,\mathrm{d}v = \int_{\mathbb{R}_3} \hat{\boldsymbol{\tau}}(\mathbf{k}) : \hat{\tilde{\boldsymbol{\varepsilon}}}^{\dagger}(\mathbf{k})\,\mathrm{d}v_{\mathbf{k}}. \tag{12.33}$$

We can solve equation (12.32) in Fourier space for $\hat{\tilde{\mathbf{u}}}(\mathbf{k})$ and use this to replace $\hat{\tilde{\boldsymbol{\varepsilon}}}(\mathbf{k})$ in equation (12.33), which can then be solved easily by using the symmetry in Fourier space.

We first note that the homogeneous part $\bar{\boldsymbol{\tau}}$ remains undetermined by equation (12.32). We can therefore only provide $\tilde{\boldsymbol{\tau}}(\mathbf{x})$, and need to carefully distinguish the two parts $\bar{\boldsymbol{\tau}}$ and $\tilde{\boldsymbol{\tau}}(\mathbf{x})$. It is clear that $\bar{\boldsymbol{\tau}}$ does not enter equation (12.31). We see in equation (12.28) that the effect of the nabla operator can be replaced in Fourier space by $\mathrm{i}\mathbf{k}$, which allows translating equation (12.32) to Fourier space,

$$(\mathbb{C}^0 : \hat{\tilde{\mathbf{u}}}(\mathbf{k}) \otimes \mathrm{i}\mathbf{k} + \hat{\tilde{\boldsymbol{\tau}}}(\mathbf{k})) \cdot \mathrm{i}\mathbf{k} = \mathbf{o}. \tag{12.34}$$

This is solved for $\hat{\tilde{\mathbf{u}}}(\mathbf{k})$,

$$\hat{\tilde{\mathbf{u}}}(\mathbf{k}) = \mathrm{i}\mathbf{A}^{-1}(\mathbf{k})\hat{\tilde{\boldsymbol{\tau}}}(\mathbf{k}) \cdot \mathbf{k} \quad \text{with } \mathbf{A}(\mathbf{k}) = \mathbf{k} \cdot \mathbb{C}^0 \cdot \mathbf{k}. \tag{12.35}$$

$\mathbf{A}(\mathbf{k})$ is the acoustic tensor. In the case of isotropy, the reference stiffness tetrad is written as

$$\mathbb{C}^0 = \lambda_0 \mathbf{I} \otimes \mathbf{I} + \mu_0(\delta_{ik}\delta_{jl} + \delta_{il}\delta_{jk})\mathbf{e}_i \otimes \mathbf{e}_j \otimes \mathbf{e}_k \otimes \mathbf{e}_l. \tag{12.36}$$

This is used to summarize the acoustic tensor

$$\mathbf{A}(\mathbf{k}) = (\lambda_0 + \mu_0)\mathbf{k} \otimes \mathbf{k} + \mu_0(\mathbf{k} \cdot \mathbf{k})\mathbf{I} \tag{12.37}$$

$$= k^2(\lambda_0 + 2\mu_0)\mathbf{k}^* \otimes \mathbf{k}^* + k^2\mu_0(\mathbf{I} - \mathbf{k}^* \otimes \mathbf{k}^*), \tag{12.38}$$

$$\mathbf{A}^{-1}(\mathbf{k}) = \frac{1}{k^2(\lambda_0 + 2\mu_0)}\mathbf{k}^* \otimes \mathbf{k}^* + \frac{1}{k^2\mu_0}(\mathbf{I} - \mathbf{k}^* \otimes \mathbf{k}^*) \tag{12.39}$$

$$= \frac{1}{k^2}\Big[\underbrace{\Big(\frac{1}{\lambda_0 + 2\mu_0} - \frac{1}{\mu_0}\Big)}_{\alpha}\mathbf{k}^* \otimes \mathbf{k}^* + \underbrace{\frac{1}{\mu_0}}_{\beta}\mathbf{I}\Big], \tag{12.40}$$

where the decomposition of $\mathbf{k} = k\mathbf{k}^*$ into magnitude k and direction $\mathbf{k}^*$ with $k = \sqrt{\mathbf{k} \cdot \mathbf{k}}$ is used. These expressions have already appeared in the basic laminate solution for $\mathbf{k} = \mathbf{e}_3$. We obtain $\hat{\tilde{\boldsymbol{\varepsilon}}}(\mathbf{k})$ by multiplying $\hat{\tilde{\mathbf{u}}}(\mathbf{k})$ with $i\mathbf{k}$ and insert this into the integral (12.33),

$$\int_{\mathbb{R}_3} \boldsymbol{\tau}(\mathbf{x}) : \tilde{\boldsymbol{\varepsilon}}(\mathbf{x})\,\mathrm{d}v = \int_{\mathbb{R}_3} \hat{\tilde{\boldsymbol{\tau}}}(\mathbf{k}) : (-\mathbf{k} \otimes \mathbf{A}^{-1} \otimes \mathbf{k}) : \hat{\tilde{\boldsymbol{\tau}}}(\mathbf{k})\,\mathrm{d}v_{\mathbf{k}} \tag{12.41}$$

$$= -\int_{\mathbb{R}_3} \hat{\tilde{\boldsymbol{\tau}}}(\mathbf{k}) : (\mathbf{k}^* \otimes (\alpha\mathbf{k}^* \otimes \mathbf{k}^* + \beta\mathbf{I}) \otimes \mathbf{k}^*) : \hat{\tilde{\boldsymbol{\tau}}}(\mathbf{k})\,\mathrm{d}v_{\mathbf{k}}. \tag{12.42}$$

It is noteworthy and important that the magnitude k of $\mathbf{k}$ appears only indirectly through $\hat{\tilde{\boldsymbol{\tau}}}(\mathbf{k})$ in the integral.

The integrand is spherically symmetric. Hence, we can nicely integrate by parameterizing with spherical coordinates,

$$k_1 = k \sin\theta \cos\phi, \tag{12.43}$$

$$k_2 = k \sin\theta \sin\phi, \tag{12.44}$$

$$k_3 = k \cos\theta. \tag{12.45}$$

Compared to the geographic coordinate system of the earth, the coordinate lines along a varying ϕ correspond to the parallels, and the coordinate lines when varying θ correspond to the meridians. Unlike the parallels of the earth's graticules that are measured in degrees north or south of the equator, here θ runs from the north pole at $\theta = 0$ to the south pole at $\theta = \pi$. The ordering of the coordinates is r, θ, ϕ. This choice of θ running from one pole to the other has some advantages, as we will see later in Section 13.5. In any case, this is the usual convention in mathematics. The integration over spherical coordinates requires inserting the determinant of the Jacobi[6] matrix of the coordinate change

$$J = \det(\partial k_{1,2,3}/\partial(k,\theta,\phi)) = k^2 \sin\theta, \tag{12.46}$$

the origin of which is detailed in Section 13.5. Due to the isotropy of the microstructure, $\hat{\tilde{\boldsymbol{\tau}}}(\mathbf{k})$ depends only on k but not on ϕ or θ. With this, the integral (12.42) becomes

6 Carl Gustav Jacob Jacobi, 1804–1951 (https://en.wikipedia.org/wiki/Carl_Gustav_Jacob_Jacobi).

$$\int_{\mathbb{R}_3} \boldsymbol{\tau}(\mathbf{x}) : \tilde{\boldsymbol{\varepsilon}}(\mathbf{x})\,\mathrm{d}v = -\int_0^\infty k^2 \hat{\tilde{\boldsymbol{\tau}}}(k) \otimes \hat{\tilde{\boldsymbol{\tau}}}(k) :: \int_0^{2\pi}\int_0^{\pi} \mathbf{k}^* \otimes (\alpha\mathbf{k}^* \otimes \mathbf{k}^* + \beta\mathbf{I}) \otimes \mathbf{k}^* \sin\theta\,\mathrm{d}\theta\,\mathrm{d}\phi\,\mathrm{d}k. \tag{12.47}$$

It is understandable that the integration of $\mathbf{k}^* \otimes \mathbf{I} \otimes \mathbf{k}^*$ and $\mathbf{k}^* \otimes \mathbf{k}^* \otimes \mathbf{k}^* \otimes \mathbf{k}^*$ over a unit sphere results in isotropic tensors, and involves only the laborious integration of products of sine and cosine functions. The result is

$$\int_0^{2\pi}\int_0^{\pi} \mathbf{k}^* \otimes \mathbf{k}^* \otimes \mathbf{k}^* \otimes \mathbf{k}^* \cos\theta\,\mathrm{d}\theta\,\mathrm{d}\phi = \frac{4\pi}{15}(\delta_{ij}\delta_{kl} + \delta_{ik}\delta_{jl} + \delta_{il}\delta_{jk})\mathbf{e}_i \otimes \mathbf{e}_j \otimes \mathbf{e}_k \otimes \mathbf{e}_l \tag{12.48}$$

$$\int_0^{2\pi}\int_0^{\pi} \mathbf{k}^* \otimes \mathbf{I} \otimes \mathbf{k}^* \cos\theta\,\mathrm{d}\theta\,\mathrm{d}\phi = \frac{4\pi}{3}\delta_{il}\delta_{jk}\mathbf{e}_i \otimes \mathbf{e}_j \otimes \mathbf{e}_k \otimes \mathbf{e}_l, \tag{12.49}$$

which can be computed with the aid of the Mathematica notebook in Listing 12.1.

Listing 12.1: Mathematica notebook (https://gitlab.com/gluegerainer/listings-homogenization-methods/-/blob/main/Listing-15_en.nb) for integrating $\mathbf{k}^* \otimes \mathbf{k}^* \otimes \mathbf{k}^* \otimes \mathbf{k}^*$ and $\mathbf{k}^* \otimes \mathbf{I} \otimes \mathbf{k}^*$ over the unit sphere.

```
k = {Cos[phi] Sin[theta], Sin[phi] Sin[theta], Cos[theta]};
id = IdentityMatrix[3];
kkkk = Integrate[ Sin[theta] TensorProduct[k, k, k, k], {theta, 0, Pi}, {phi, 0, 2 Pi}]
kIk = Integrate[ Sin[theta] TensorProduct[k, id, k], {theta, 0, Pi}, {phi, 0, 2 Pi}]
(* Verification: Should yield only zeros *)
kkkk - 4 Pi/15 Table[ id[[i, j]] id[[k, l]] + id[[i, k]] id[[j, l]] +  id[[i, l]] id[[k, j]], {i
    , 3}, {j, 3}, {k, 3}, {l, 3}]
kIk - 4 Pi/3 Table[ id[[i, l]] id[[k, j]], {i, 3}, {j, 3}, {k, 3}, {l, 3}]
```

The remaining integral is

$$\int_{\mathbb{R}_3} \tilde{\boldsymbol{\tau}}(\mathbf{x}) : \tilde{\boldsymbol{\varepsilon}}(\mathbf{x})\,\mathrm{d}v = -\int_0^\infty \hat{\tilde{\boldsymbol{\tau}}}(k) : \mathbb{W} : \hat{\tilde{\boldsymbol{\tau}}}(k) 4\pi k^2\,\mathrm{d}k, \tag{12.50}$$

with

$$\mathbb{W} = \left[\frac{\alpha}{15}(\delta_{ij}\delta_{kl} + \delta_{ik}\delta_{jl} + \delta_{il}\delta_{jk}) + \frac{\beta}{3}\delta_{il}\delta_{jk}\right]\mathbf{e}_i \otimes \mathbf{e}_j \otimes \mathbf{e}_k \otimes \mathbf{e}_l. \tag{12.51}$$

$\mathbb{W}$ is the interaction tensor that couples the volumetric strain with the deviatoric strain. Unlike the Voigt and Reuss bounds, for which we have separate mixture rules for K and G, the compression and shear moduli interact in the case of the Hashin–Shtrikman bounds.

We see that the remaining integral over k contains the infinitesimally thin spherical shell volume $4\pi k^2\,\mathrm{d}k = \mathrm{d}V$. Since the integral is over the entire Fourier space, we can convert it into an integral over real space using Plancherel's theorem,

$$\int_{\mathbb{R}_3} \tilde{\boldsymbol{\tau}}(\mathbf{x}) : \tilde{\boldsymbol{\varepsilon}}(\mathbf{x})\,\mathrm{d}v = -\int_{\mathbb{R}_3} \tilde{\boldsymbol{\tau}}(\mathbf{x}) : \mathbb{W} : \tilde{\boldsymbol{\tau}}(\mathbf{x})\,\mathrm{d}v. \tag{12.52}$$

Finally, we have replaced $\tilde{\boldsymbol{\varepsilon}}(\mathbf{x})$ by $\mathbb{W} : \tilde{\boldsymbol{\tau}}(\mathbf{x})$, and one can evaluate or minimize the Hashin–Shtrikman functional over the coefficients in some approach for $\boldsymbol{\tau}(\mathbf{x})$. Since all stiffnesses are isotropic, the decomposition of $\mathbb{W}$ w. r. t. the isotropic projectors is feasible. To do so, we first generate the subsymmetries,

$$\mathbb{W}^{\mathrm{sym}} = \frac{1}{4}(W_{ijkl} + W_{ijlk} + W_{jikl} + W_{jilk})\mathbf{e}_i \otimes \mathbf{e}_j \otimes \mathbf{e}_k \otimes \mathbf{e}_l. \tag{12.53}$$

Without loss of generality, we can proceed with the symmetrized $\mathbb{W}^{\mathrm{sym}}$, dropping the explicit sym-superscript. A full scalar contraction with the isotropic projectors gives the components

$$w_1 = \mathbb{W}^{\mathrm{sym}} :: \mathbb{P}_{I1} = \frac{\alpha + \beta}{3}, \tag{12.54}$$

$$w_2 = \mathbb{W}^{\mathrm{sym}} :: \mathbb{P}_{I2}/5 = \frac{2\alpha + 5\beta}{15} \tag{12.55}$$

w. r. t. the projector representation

$$\mathbb{W}^{\mathrm{sym}} = w_1\mathbb{P}_{I1} + w_2\mathbb{P}_{I2}, \tag{12.56}$$

where we need to consider the normalizing factor 1/5 of the second isotropic projector; see Section 2.4. This computation can also be passed on to a computer algebra system; see Listing 12.2.

Listing 12.2: Mathematica notebook (https://gitlab.com/gluegerainer/listings-homogenization-methods/-/blob/main/Listing-15_en.nb) for finding the projector representation of $\mathbb{W}^{\mathrm{sym}}$, complementary to Listing 12.1.

```
W = alpha/4/Pi kkkk + beta /4/Pi kIk;
Wsym = Table[ W[[i, j, k, l]] + W[[i, j, l, k]] + W[[j, i, k, l]] + W[[j, i, l, k]], {i, 1, 3},
    {j, 1, 3}, {k, 1, 3}, {l, 1, 3}]/4 // FullSimplify;
P1 = Table[ id[[i, j]] id[[k, l]], {i, 1, 3}, {j, 1, 3}, {k, 1, 3}, {l, 1, 3}]/3;
P2 = Table[ id[[i, k]] id[[j, l]] + id[[i, l]] id[[j, k]], {i, 1, 3}, {j, 1, 3}, {k, 1, 3}, {l,
    1, 3}]/2 - P1;
w1 = Sum[Wsym[[i, j, k, l]] P1[[i, j, k, l]], {i, 1, 3}, {j, 1, 3}, {k, 1, 3}, {l, 1, 3}] //
    FullSimplify
w2 = Sum[Wsym[[i, j, k, l]] P2[[i, j, k, l]], {i, 1, 3}, {j, 1, 3}, {k, 1, 3}, {l, 1, 3}]/5 //
    FullSimplify
(* Verification: Should yield only zeros *)
Wsym - w1 P1 - w2 P2 // FullSimplify
```

Concluding, we rewrite the symmetrized $\mathbb{W}$ in terms of K and G with α and β according to equation (12.40) and with $\lambda = K - 2G/3$ and $\mu = G$,

$$\mathbb{W}^{\mathrm{sym}} = \underbrace{\frac{1}{3K + 4G}}_{w_1}\mathbb{P}_{I1} + \underbrace{\frac{3(K + 2G)}{5(3KG + 4G^2)}}_{w_2}\mathbb{P}_{I2}. \tag{12.57}$$

In this form, one can see the interaction part of W nicely:
- The shear modulus G, i. e., the resistance against volume preserving deformations, appears in front of the first isotropic projector, which extracts the volumetric strains when applied to a strain tensor.
- The compression modulus K, i. e., the resistance against volume changes, appears in front of the second isotropic projector, which extracts the volume preserving (deviatoric) deformations when applied to a strain tensor.

12.5.2 Heat conduction and diffusion

Instead of requiring the vanishing divergence of the stress tensor $\boldsymbol{\sigma} \cdot \nabla = \mathbf{o}$ due to the preservation of momentum without body forces, we require the vanishing divergence of a heat or concentration flux $\mathbf{q} \cdot \nabla = 0$ due to the conservation of energy or mass without source terms. We can mostly transfer the symbolic notation from the previous section. We merely need to interpret

$$\boldsymbol{\tau} = \left(\mathbf{L}(\mathbf{x}) - \mathbf{L}^0\right) \cdot \mathbf{g}(\mathbf{x}) \tag{12.58}$$

with the temperature or concentration gradient $\mathbf{g} = \nabla T$. Here, we choose the temperature T for heat conduction. The diffusion problem is completely analogous. The balance $\mathbf{q} \cdot \nabla = 0$ is

$$\left(\mathbf{L}^0 \cdot \tilde{\mathbf{g}} + \boldsymbol{\tau}\right) \cdot \nabla = 0. \tag{12.59}$$

The Fourier transformation is

$$\left(\mathbf{L}^0 \cdot \mathrm{i}\mathbf{k}\hat{\tilde{T}} + \mathrm{i}\hat{\boldsymbol{\tau}}\right) \cdot \mathrm{i}\mathbf{k} = 0 \tag{12.60}$$

$$-\hat{\tilde{T}}\mathbf{L}^0 \cdot\cdot \mathbf{k} \otimes \mathbf{k} + \mathrm{i}\hat{\boldsymbol{\tau}} \cdot \mathbf{k} = 0. \tag{12.61}$$

We can solve for $\hat{\tilde{T}}$,

$$\hat{\tilde{T}} = \frac{\mathrm{i}\hat{\boldsymbol{\tau}} \cdot \mathbf{k}}{\mathbf{L}^0 \cdot\cdot \mathbf{k} \otimes \mathbf{k}}, \tag{12.62}$$

which becomes upon choosing $\mathbf{L}^0 = l_0 \mathbf{I}$

$$\hat{\tilde{T}} = \frac{\mathrm{i}\hat{\boldsymbol{\tau}} \cdot \mathbf{k}}{l_0 \mathbf{k} \cdot \mathbf{k}}. \tag{12.63}$$

This can be used to write the Hashin–Shtrikman functional that is over the real space as an integral over Fourier space by using Plancherel's theorem,

$$\int_{\mathbb{R}_{\text{real}}} \boldsymbol{\tau} \cdot \tilde{\mathbf{g}} \, \mathrm{d}v = \int_{\mathbb{R}_{\text{Fourier}}} \hat{\boldsymbol{\tau}} \cdot \hat{\tilde{\mathbf{g}}}^{\dagger} \, \mathrm{d}v_{\mathbf{k}} \tag{12.64}$$

$$= l_0^{-1} \int_{\mathbb{R}_{\text{Fourier}}} \hat{\boldsymbol{\tau}} \otimes \hat{\boldsymbol{\tau}} : \mathbf{k}^* \otimes \mathbf{k}^* \, \mathrm{d}v_{\mathbf{k}}, \tag{12.65}$$

with $\mathbf{k}^* = \mathbf{k}/k$, $k = |\mathbf{k}|$.

Isotropy in $\mathbb{R}_2$ – transversal isotropy

We parameterize the plane with polar coordinates k and ϕ,

$$k_1 = k \cos\phi \tag{12.66}$$

$$k_2 = k \sin\phi, \tag{12.67}$$

the determinant of the Jacobian matrix is $J = k = |\mathbf{k}|$. The integral becomes

$$\int_{\mathbb{R}_2} \boldsymbol{\tau} \cdot \tilde{\mathbf{g}} \, \mathrm{d}v = l_0^{-1} \int_0^{\infty} k \hat{\boldsymbol{\tau}} \otimes \hat{\boldsymbol{\tau}} : \int_0^{2\pi} \mathbf{k}^* \otimes \mathbf{k}^* \, \mathrm{d}\phi \, \mathrm{d}k. \tag{12.68}$$

The inner integral is

$$\int_0^{2\pi} \mathbf{k}^* \otimes \mathbf{k}^* \, \mathrm{d}\phi = \int_0^{2\pi} \begin{bmatrix} \cos^2\phi & \cos\phi \, \sin\phi \\ \cos\phi \, \sin\phi & \sin^2\phi \end{bmatrix} \mathbf{e}_{1,2} \otimes \mathbf{e}_{1,2} \, \mathrm{d}\phi \tag{12.69}$$

$$= \begin{bmatrix} \pi & 0 \\ 0 & \pi \end{bmatrix} \mathbf{e}_{1,2} \otimes \mathbf{e}_{1,2} = \frac{2\pi}{2} \begin{bmatrix} 1 & 0 \\ 0 & 1 \end{bmatrix} \mathbf{e}_{1,2} \otimes \mathbf{e}_{1,2}. \tag{12.70}$$

The remaining integral is $\int_0^{\infty} \cdot 2\pi \, \mathrm{d}k$ is apparently the integral over the entire plane in Fourier space with the area of an infinitesimal annulus $2\pi k \, \mathrm{d}k$, which can be converted into a real space integral by Plancherel's theorem

$$\int_{\mathbb{R}_2} \boldsymbol{\tau} \cdot \tilde{\mathbf{g}} \, \mathrm{d}v = l_0^{-1} \int_0^{\infty} \hat{\boldsymbol{\tau}} \otimes \hat{\boldsymbol{\tau}} : \frac{1}{2} \mathbf{I} \, 2\pi k \, \mathrm{d}k \tag{12.71}$$

$$= \int_{\mathbb{R}_2} \boldsymbol{\tau} \otimes \boldsymbol{\tau} : \mathbf{W}_2 \, \mathrm{d}v \tag{12.72}$$

with $\mathbf{W}_2 = (2l_0)^{-1}\mathbf{I}$, where $\mathbf{I} = \mathbf{e}_1 \otimes \mathbf{e}_1 + \mathbf{e}_2 \otimes \mathbf{e}_2$.

Isotropy in $\mathbb{R}_3$

We parameterize the Fourier space with spherical coordinates,

$$k_1 = k \sin\theta \cos\phi \tag{12.73}$$

$$k_2 = k \sin\theta \sin\phi \tag{12.74}$$

$$k_3 = k \cos\theta, \tag{12.75}$$

the determinant of the Jacobian matrix is $J = k^2 \sin\theta$. The integral becomes

$$\int_{\mathbb{R}_3} \boldsymbol{\tau} \cdot \tilde{\mathbf{g}} \, \mathrm{d}v = l_0^{-1} \int_0^\infty k^2 \hat{\boldsymbol{\tau}} \otimes \hat{\boldsymbol{\tau}} : \int_0^{2\pi} \int_0^{\pi} \mathbf{k}^* \otimes \mathbf{k}^* \sin\theta \, \mathrm{d}\theta \, \mathrm{d}\phi \, \mathrm{d}k. \tag{12.76}$$

The inner integral is

$$\int_0^{2\pi} \int_0^{\pi} \sin\theta \mathbf{k}^* \otimes \mathbf{k}^* \, \mathrm{d}\theta \, \mathrm{d}\phi = \frac{4\pi}{3} \mathbf{I}. \tag{12.77}$$

The remaining integral $\int_0^\infty \cdot 4\pi k^2 \, \mathrm{d}k$ is apparently the integral over the entire Fourier space with the infinitesimal volume of a spherical shell $4\pi k^2 \, \mathrm{d}k$, which allows applying Plancherel's theorem again to convert it into a real space integral,

$$\int_{\mathbb{R}_3} \boldsymbol{\tau} \cdot \tilde{\mathbf{g}} \, \mathrm{d}v = \int_{\mathbb{R}_3} \boldsymbol{\tau} \otimes \boldsymbol{\tau} : \mathbf{W}_3 \, \mathrm{d}v \tag{12.78}$$

with $\mathbf{W}_3 = (3l_0)^{-1} \mathbf{I}$.

12.6 Decomposition of periodic fields into three orthogonal parts

The Fourier representation allows decomposing a field algebraically into the mean part, the divergence-free part and the rotation-free part. In elastostatics and heat conduction/diffusion, the local balances, linear constitutive laws and integrability conditions are

$$\nabla \cdot \boldsymbol{\sigma} = \mathbf{o}, \quad \boldsymbol{\sigma} = \mathbb{C} : \boldsymbol{\varepsilon}, \quad \nabla \times \boldsymbol{\varepsilon} \times \nabla = \mathbf{O}, \tag{12.79}$$

$$\nabla \cdot \mathbf{q} = 0, \quad \mathbf{q} = \mathbf{L} \cdot \mathbf{g}, \quad \nabla \times \mathbf{g} = \mathbf{o}. \tag{12.80}$$

The differential constraints on $\boldsymbol{\varepsilon}$ and $\mathbf{g}$ are called compatibility conditions. They ensure that to a $\boldsymbol{\varepsilon}$ field, a displacement field $\mathbf{u}$ exists such that $\boldsymbol{\varepsilon} = \operatorname{sym}(\mathbf{u} \otimes \nabla)$. The same holds for the temperature gradient $\mathbf{g} = \nabla T$. Up to now, the integrability conditions have played no role. Their appearance implies an additional equation that can be used for deriving approximations of the effective properties.

We consider the heat conduction problem on a periodically repeatable unit cell, i. e., the fields $\{\mathbf{q}, \mathbf{L}, \mathbf{g}\}$ are periodic. In the following derivation, the notation can well be interpreted more generally for the fields $\{\boldsymbol{\sigma}, \mathbb{C}, \boldsymbol{\varepsilon}\}$ of elastostatics.

Let us take the divergence and the rotation of a vector field in the Fourier representation,

$$\mathbf{f}(\mathbf{x}) \cdot \nabla = \sum_{\forall \mathbf{k}} \mathrm{i}\, \mathrm{e}^{\mathrm{i}\mathbf{x}\cdot\mathbf{k}} (\hat{\mathbf{f}}_{\mathbf{k}} \cdot \mathbf{k}). \tag{12.81}$$

This can be expanded by $\mathbf{k}$ and normalized,

$$\mathbf{f}_{\cdot\nabla}(\mathbf{x}) := \sum_{\forall \mathbf{k}} \mathrm{i}\, \mathrm{e}^{\mathrm{i}\mathbf{x}\cdot\mathbf{k}} \mathbf{K} \hat{\mathbf{f}}_{\mathbf{k}} \tag{12.82}$$

with

$$\mathbf{K} = (\mathbf{k} \otimes \mathbf{k})/(\mathbf{k} \cdot \mathbf{k}) = \mathbf{k}^* \otimes \mathbf{k}^*, \tag{12.83}$$

$$\mathbf{k}^* = \mathbf{k}/k \quad \text{with } k = |\mathbf{k}| = \sqrt{\mathbf{k} \cdot \mathbf{k}}, \tag{12.84}$$

with the decomposition of $\mathbf{k}$ into magnitude and direction $\mathbf{k} = k\mathbf{k}^*$. k is the wave number and $\mathbf{k}^*$ the direction of the plane wave. An analogous calculation for the rotation $\times\nabla$ yields

$$\mathbf{f}(\mathbf{x}) \times \nabla = \sum_{\forall \mathbf{k}} \mathrm{i}\, \mathrm{e}^{\mathrm{i}\mathbf{x}\cdot\mathbf{k}} (\hat{\mathbf{f}}_{\mathbf{k}} \times \mathbf{k}). \tag{12.85}$$

One can see that in the case of the rotation, the divergence part is filtered out and vice versa: The cross product is zero for parallel vectors, i. e., $\mathbf{k} \times \mathbf{k} = \mathbf{o}$. Therefore, we can expand the last equation with $\mathbf{I} - \mathbf{K}$,

$$\mathbf{f}_{\times\nabla}(\mathbf{x}) := \sum_{\forall \mathbf{k}} \mathrm{i}\, \mathrm{e}^{\mathrm{i}\mathbf{x}\cdot\mathbf{k}} (\mathbf{I} - \mathbf{K}) \hat{\mathbf{f}}_{\mathbf{k}}. \tag{12.86}$$

The Fourier component to $\mathbf{k} = \mathbf{o}$ represents a homogeneous (constant) vector field, which is divergence as well as rotation-free, and we could join it with any of the other parts. But since it corresponds to the field's mean value $\bar{\mathbf{f}}$, we keep it as a separate part. Hence, the field $\mathbf{f}(\mathbf{x})$ is decomposed into three parts:

$$\mathbf{f}(\mathbf{x}) = \underbrace{\hat{\mathbf{f}}_{\mathbf{o}}}_{\text{homogeneous part } \bar{\mathbf{f}}} + \underbrace{\sum_{\forall \mathbf{k} \neq \mathbf{o}} \mathrm{e}^{\mathrm{i}\mathbf{x}\cdot\mathbf{k}} (\mathbf{I} - \mathbf{K}) \hat{\mathbf{f}}_{\mathbf{k}}}_{\text{divergence free part } \tilde{\mathbf{f}}_{\times\nabla}(\mathbf{x})} + \underbrace{\sum_{\forall \mathbf{k} \neq \mathbf{o}} \mathrm{e}^{\mathrm{i}\mathbf{x}\cdot\mathbf{k}} \mathbf{K} \hat{\mathbf{f}}_{\mathbf{k}}}_{\text{rotation free part } \tilde{\mathbf{f}}_{\cdot\nabla}(\mathbf{x})} . \tag{12.87}$$

It is due to the projectors $\mathbf{K}$ and $\mathbf{I} - \mathbf{K}$ apparent that the nonhomogenous parts are perpendicular to each other. Hence, we apply in the Fourier representation the projectors $\mathbf{K}$ and $\mathbf{I} - \mathbf{K}$ to filter out the rotation-free and the divergence-free parts algebraically, *without the imaginary unit*, which is basically responsible for the derivative.

In real space, the orthogonality is apparent as well. Applying the rotation produces a divergence-free field and vice versa,

$$(\mathbf{f}(\mathbf{x}) \times \nabla) \cdot \nabla = f_{i,jk}\varepsilon_{ijk} = 0. \tag{12.88}$$

This product vanishes because the permutation symbol $\varepsilon_{ijk} = -\varepsilon_{ikj}$ is antisymmetric and the symmetry of second derivatives $f_{i,jk} = f_{i,kj}$. But it is much harder to separate these two parts in real space. In Fourier space, we write down algebraic operators that filter these parts,

$$\boldsymbol{\Gamma}_{0}\{\mathbf{f}(\mathbf{x})\} = \bar{\mathbf{f}} = \hat{\mathbf{f}}_{\mathbf{o}} \tag{12.89}$$

$$\boldsymbol{\Gamma}_{.\nabla}\{\mathbf{f}(\mathbf{x})\} = \mathbf{f}(\mathbf{x}) - \bar{\mathbf{f}} - \tilde{\mathbf{f}}_{\times\nabla}(\mathbf{x}) = \tilde{\mathbf{f}}_{.\nabla}(\mathbf{x}) = \sum_{\forall\mathbf{k}\neq\mathbf{o}} \mathrm{e}^{\mathrm{i}\mathbf{x}\cdot\mathbf{k}}\mathbf{K}\hat{\mathbf{f}}_{\mathbf{k}} \tag{12.90}$$

$$\boldsymbol{\Gamma}_{\times\nabla}\{\mathbf{f}(\mathbf{x})\} = \mathbf{f}(\mathbf{x}) - \bar{\mathbf{f}} - \tilde{\mathbf{f}}_{.\nabla}(\mathbf{x}) = \tilde{\mathbf{f}}_{\times\nabla}(\mathbf{x}) = \sum_{\forall\mathbf{k}\neq\mathbf{o}} \mathrm{e}^{\mathrm{i}\mathbf{x}\cdot\mathbf{k}}(\mathbf{I} - \mathbf{K})\hat{\mathbf{f}}_{\mathbf{k}}. \tag{12.91}$$

The parts $\tilde{\mathbf{f}}_{\times\nabla}(\mathbf{x})$ and $\tilde{\mathbf{f}}_{.\nabla}(\mathbf{x})$ are the fluctuating parts of the fields $\mathbf{f}(x)$, which do not satisfy the differential constraints $\mathbf{f}(\mathbf{x}) \cdot \nabla = 0$ and $\mathbf{f}(\mathbf{x}) \times \nabla = \mathbf{o}$. The subscript at $\boldsymbol{\Gamma}$ shows which part of the field $\mathbf{f}$ remains when the operator $\boldsymbol{\Gamma}$ is applied. This is the Stokes[7]–Helmholtz[8] decomposition, also known as the fundamental theorem of vector analysis. With respect to the scalar product, these fields are orthogonal,

$$[\tilde{\mathbf{f}}_{.\nabla}(\mathbf{x}), \tilde{\mathbf{f}}^{\dagger}_{\times\nabla}(\mathbf{x})] = \frac{1}{V}\int_{\Omega} \tilde{\mathbf{f}}_{.\nabla}(\mathbf{x}) \cdot \tilde{\mathbf{f}}^{\dagger}_{\times\nabla}(\mathbf{x})\,\mathrm{d}v \tag{12.92}$$

$$= \frac{1}{V}\int_{\Omega} \sum_{\forall\mathbf{k}_1\neq\mathbf{o}} \mathrm{e}^{\mathrm{i}\mathbf{k}_1\cdot\mathbf{x}}(\mathbf{k}_1^* \otimes \mathbf{k}_1^*)\hat{\mathbf{f}}_{\mathbf{k}_1} \cdot \sum_{\forall\mathbf{k}_2\neq\mathbf{o}} \mathrm{e}^{-\mathrm{i}\mathbf{k}_2\cdot\mathbf{x}}(\mathbf{I} - \mathbf{k}_2^* \otimes \mathbf{k}_2^*)\hat{\mathbf{f}}_{\mathbf{k}_2}\,\mathrm{d}v. \tag{12.93}$$

The product of the exponentials is only nonzero for $\mathbf{k}_1 = \mathbf{k}_2 = \mathbf{k}$ (see Section 12.1), such that we can replace the exponential functions by the factor 1 for this case,

$$[\tilde{\mathbf{f}}_{.\nabla}(\mathbf{x}), \tilde{\mathbf{f}}^{\dagger}_{\times\nabla}(\mathbf{x})] = \frac{1}{V}\int_{\Omega} \sum_{\forall\mathbf{k}\neq\mathbf{o}} \hat{\mathbf{f}}_{\mathbf{k}} \cdot \underbrace{(\mathbf{k}^* \otimes \mathbf{k}^*)(\mathbf{I} - \mathbf{k}^* \otimes \mathbf{k}^*)}_{\mathbf{o}}\,\hat{\mathbf{f}}_{\mathbf{k}}\,\mathrm{d}v = 0. \tag{12.94}$$

On the other hand, in the case of $\mathbf{k}_1 = \mathbf{k}_2 = \mathbf{k}$, the product of the projectors in the integral vanishes, which renders the entire expression equal to zero.

The $\boldsymbol{\Gamma}$-operators cannot be applied as if they were matrices, and just like the nabla operator, they are not associative. Therefore, we specify the argument to which they are applied with curly braces.

We have denoted the decomposition for vector field problems like diffusion or heat conduction, but the calculation can be transferred with minor changes to elastostatics, where the decomposition becomes

7 George Gabriel Stokes, 1819–1903 (https://en.wikipedia.org/wiki/George_Gabriel_Stokes).

8 Hermann von Helmholtz, 1821–1894 (https://en.wikipedia.org/wiki/Hermann_von_Helmholtz).

$$\mathbf{A}(\mathbf{x}) = \underbrace{\hat{\mathbf{A}}_{\mathbf{o}}}_{\text{hom. part}} + \underbrace{\sum_{\forall \mathbf{k} \neq \mathbf{o}} e^{i\mathbf{x}\cdot\mathbf{k}} (\mathbb{I} - \mathbf{K} \otimes \mathbf{K}) : \hat{\mathbf{A}}_{\mathbf{k}}}_{\text{divergence free fluctuating part } \tilde{\mathbf{A}}_{\times\nabla}(\mathbf{x})} + \underbrace{\sum_{\forall \mathbf{k} \neq \mathbf{o}} e^{i\mathbf{x}\cdot\mathbf{k}} \mathbf{K} \otimes \mathbf{K} : \hat{\mathbf{A}}_{\mathbf{k}}}_{\text{rotation free fluctuating part } \tilde{\mathbf{A}}_{\cdot\nabla}(\mathbf{x})} \tag{12.95}$$

for a symmetric second-order tensor field $\mathbf{A}(\mathbf{x})$. Milton (2002) is w. r. t. this Fourier method more detailed.

12.6.1 Determination of the effective properties

Let us consider the polarization problem (see Section 8.3) for heat conduction with a reference conductivity $\mathbf{L}_0$. All quantities except $\mathbf{L}_0$, the mean values $\overline{\mathbf{q}}$ and $\overline{\mathbf{g}}$ and the effective conductivity $\mathbf{L}^*$ depend on the location $\mathbf{x}$ in real space or $\mathbf{k}$ in Fourier space, but we do not write out this dependency explicitly for brevity,

$$\mathbf{p} = (\mathbf{L} - \mathbf{L}_0)\mathbf{g} \tag{12.96}$$

$$= \mathbf{q} - \mathbf{L}_0\mathbf{g}, \tag{12.97}$$

with $\mathbf{q} = \mathbf{L}\mathbf{g}$. Taking the average gives

$$\overline{\mathbf{p}} = \overline{\mathbf{q}} - \mathbf{L}_0\overline{\mathbf{g}}. \tag{12.98}$$

We have the three Stokes–Helmholtz-projectors $\mathbf{\Gamma}_{(\cdot)}$, which extract the components of the Stokes–Helmholtz decomposition and the constant (homogeneous) part of a field. They are differential operators in real space and simple, algebraic projectors in Fourier space. $\mathbf{\Gamma}_0$ extracts the mean value of a field, and $\mathbf{\Gamma}_{\times\nabla}$ and $\mathbf{\Gamma}_{\cdot\nabla}$ extract the divergence-free and rotation-free parts of the fluctuating part of the field, respectively. For simplicity, we use $\mathbf{L}_0 = l_0\mathbf{I}$. With these preparations, we derive an expression for $\overline{\mathbf{p}}$ that is linear in $\overline{\mathbf{g}}$. A comparison of coefficients gives then $\mathbf{L}^*$.

We firstly apply $\mathbf{\Gamma}_{\cdot\nabla}\{\mathbf{L}_0^{-1}\{\cdot\}\}$ to equation (12.97),

$$\mathbf{\Gamma}_{\cdot\nabla}\{\mathbf{L}_0^{-1}\mathbf{p}\} = \mathbf{\Gamma}_{\cdot\nabla}\{\mathbf{L}_0^{-1}\mathbf{q}\} - \mathbf{\Gamma}_{\cdot\nabla}\{\mathbf{g}\}. \tag{12.99}$$

Since $\mathbf{g} = \overline{\mathbf{g}} + \tilde{\mathbf{g}}_{\cdot\nabla}$ is rotation-free (equation (12.80)), this just removes the homogeneous part from $\mathbf{g}$. Since $\mathbf{q}$ is divergence-free and $\mathbf{L}_0 = l_0\mathbf{I}$ is a constant multiple of the identity tensor, $\mathbf{L}_0^{-1}\mathbf{q} = \mathbf{q}/l_0$ is as well divergence-free, such that $\mathbf{\Gamma}_{\cdot\nabla}\{\mathbf{q}\}/l_0 = \mathbf{o}$ holds. We obtain

$$\mathbf{\Gamma}_{\cdot\nabla}\{\mathbf{L}_0^{-1}\mathbf{p}\} = -\tilde{\mathbf{g}} = -\mathbf{g} + \overline{\mathbf{g}}. \tag{12.100}$$

This can be solved for $\mathbf{g} = \overline{\mathbf{g}} - \mathbf{\Gamma}\{\mathbf{p}\}$, with the abbreviations $\mathbf{\Gamma}\{\cdot\} = \mathbf{\Gamma}_{\cdot\nabla}\{\mathbf{L}_0^{-1}\{\cdot\}\}$ and $\Delta\mathbf{L} = \mathbf{L} - \mathbf{L}_0$. Inserting $\mathbf{g}$ back into the starting equation (12.96) results in

$$\mathbf{p} = \Delta\mathbf{L}\overline{\mathbf{g}} - \Delta\mathbf{L}\mathbf{\Gamma}\{\mathbf{p}\}. \tag{12.101}$$

This can be solved for **p**,

$$(\mathbf{I} + \Delta\mathbf{L}\boldsymbol{\Gamma})\{\mathbf{p}\} = \Delta\mathbf{L}\overline{\mathbf{g}} \tag{12.102}$$

$$\mathbf{p} = (\mathbf{I} + \Delta\mathbf{L}\boldsymbol{\Gamma})^{-1}\{\Delta\mathbf{L}\overline{\mathbf{g}}\}. \tag{12.103}$$

Note that the inverse is the reverse action of the operator. Since l_0 can be chosen freely, the invertibility can, in general, be ensured. We next apply $\boldsymbol{\Gamma}_0\{\mathbf{p}\} = \overline{\mathbf{p}}$ to the latter equation,

$$\overline{\mathbf{p}} = \boldsymbol{\Gamma}_0\{(\mathbf{I} + \Delta\mathbf{L}\boldsymbol{\Gamma})^{-1}\{\Delta\mathbf{L}\overline{\mathbf{g}}\}\}. \tag{12.104}$$

We can factor out $\overline{\mathbf{g}}$. Finally, this can be equated with the result for $\overline{\mathbf{p}}$ of equation (12.98),

$$\overline{\mathbf{q}} - \mathbf{L}_0\overline{\mathbf{g}} = \boldsymbol{\Gamma}_0\{(\mathbf{I} + \Delta\mathbf{L}\boldsymbol{\Gamma})^{-1}\{\Delta\mathbf{L}\}\}\overline{\mathbf{g}} \tag{12.105}$$

$$\overline{\mathbf{q}} = (\mathbf{L}_0 + \boldsymbol{\Gamma}_0\{(\mathbf{I} + \Delta\mathbf{L}\boldsymbol{\Gamma})^{-1}\{\Delta\mathbf{L}\}\})\overline{\mathbf{g}}. \tag{12.106}$$

This enables us to identify $\mathbf{L}^*$ in $\overline{\mathbf{q}} = \mathbf{L}^*\overline{\mathbf{g}}$,

$$\mathbf{L}^* = \mathbf{L}_0 + \boldsymbol{\Gamma}_0\{(\mathbf{I} + \Delta\mathbf{L}\boldsymbol{\Gamma})^{-1}\{\Delta\mathbf{L}\}\} \tag{12.107}$$

$$= \mathbf{L}_0 + \langle\, \underbrace{(\mathbf{I} + \Delta\mathbf{L}\boldsymbol{\Gamma})^{-1}\{\Delta\mathbf{L}\}}_{\mathbf{L}_\star} \,\rangle. \tag{12.108}$$

Suppose we know $\mathbf{L}_\star$, then we can easily determine $\mathbf{L}^*$. The bracket is an operator that acts on $\Delta\mathbf{L}$. This becomes clearer when one tries to determine the correction term $\mathbf{L}_\star$. Since we have $\boldsymbol{\Gamma}$ only in Fourier space, we need to represent all involved fields as Fourier series:

$$(\mathbf{I} + \Delta\mathbf{L}\boldsymbol{\Gamma})\{\mathbf{L}_\star\} = \Delta\mathbf{L} \tag{12.109}$$

$$\mathbf{L}_\star + \Delta\mathbf{L}\boldsymbol{\Gamma}\{\mathbf{L}_\star\} = \Delta\mathbf{L} \tag{12.110}$$

$$\sum_{\forall \mathbf{a}} e^{i\mathbf{x}\cdot\mathbf{a}}\hat{\mathbf{L}}_{\star\mathbf{a}} + \sum_{\forall \mathbf{b}} e^{i\mathbf{x}\cdot\mathbf{b}}\Delta\hat{\mathbf{L}}_{\mathbf{b}} \cdot \sum_{\forall \mathbf{k}\neq \mathbf{o}} e^{i\mathbf{x}\cdot\mathbf{k}} l_0^{-1}\mathbf{K}\cdot\hat{\mathbf{L}}_{\star\mathbf{k}} = \sum_{\forall \mathbf{c}} e^{i\mathbf{x}\cdot\mathbf{c}}\Delta\hat{\mathbf{L}}_{\mathbf{c}} \tag{12.111}$$

$$\sum_{\forall \mathbf{a}} e^{i\mathbf{x}\cdot\mathbf{a}}\hat{\mathbf{L}}_{\star\mathbf{a}} + \sum_{\forall \mathbf{b},\forall \mathbf{k}\neq \mathbf{o}} e^{i\mathbf{x}\cdot(\mathbf{b}+\mathbf{k})}\Delta\hat{\mathbf{L}}_{\mathbf{b}} \cdot l_0^{-1}\mathbf{K}\cdot\hat{\mathbf{L}}_{\star\mathbf{k}} = \sum_{\forall \mathbf{c}} e^{i\mathbf{x}\cdot\mathbf{c}}\Delta\hat{\mathbf{L}}_{\mathbf{c}} \tag{12.112}$$

The second summand is a so-called convolution. Due to the vectorial sum, $\mathbf{b}+\mathbf{k}$ Fourier coefficients with higher frequencies than in the initial series representation appear (Figure 12.3). This is no issue from a purely mathematical point of view for infinite series due to $\infty + \infty = \infty$. But also, for us, there is no issue: discrete periodic fields in real space are also periodic and discrete in Fourier space. Therefore, we can take the frequencies that are outside the initial Fourier representation modulo d, where d is the order of the initial Fourier representation. A numerical example is given in Section 12.6.3.

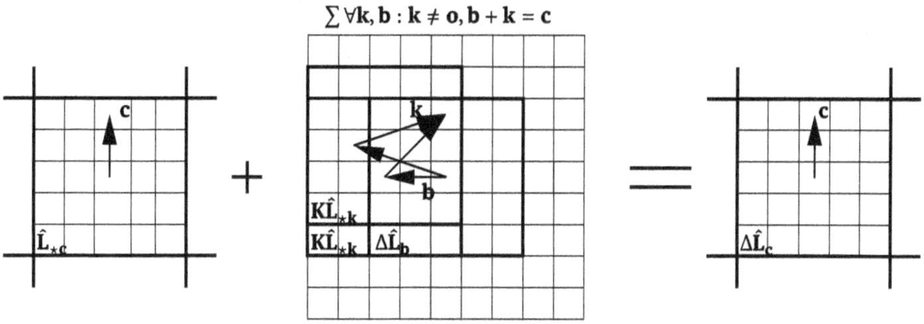

Figure 12.3: Sketch of the appearance of higher order Fourier coefficients due to the convolution equation (12.112). The initial discretization of a 2D function is with 5 × 5 Fourier coefficients. The Fourier coefficients that belong to the higher order bases $\mathrm{e}^{\mathrm{i}\mathbf{x}\cdot(\mathbf{b}+\mathbf{k})}$ that appear due to the sum $\mathbf{b} + \mathbf{k}$ correspond to the periodic repetition of the 5 × 5 cell.

We proceed with equation (12.112). A comparison of coefficients w. r. t. the common function basis $\mathrm{e}^{\mathrm{i}\mathbf{c}\cdot\mathbf{x}}$ gives

$$\hat{\mathbf{L}}_{\star\mathbf{c}} + \sum_{\forall\mathbf{k}\neq\mathbf{o}} \Delta\hat{\mathbf{L}}_{\mathbf{c}-\mathbf{k}} \cdot l_0^{-1}\mathbf{K} \cdot \hat{\mathbf{L}}_{\star\mathbf{k}} = \Delta\hat{\mathbf{L}}_{\mathbf{c}} \quad \forall\mathbf{c}. \tag{12.113}$$

This is a linear system for the unknown Fourier coefficients $\hat{\mathbf{L}}_{\star\mathbf{c}}$. We need due to the volume average (see equation (12.108)) not the entire solution, but only one solution component, namely $\hat{\mathbf{L}}_{\star\mathbf{o}}$ for $\mathbf{k} = \mathbf{o}$.

In a similar manner, one can derive other equations for $\mathbf{L}^*$ by using $\mathbf{\Gamma}_{\times\nabla}$. Moreover, instead of the linear system, one can give series expansions for $\mathbf{L}^*$. Equation (12.108) can be expanded as a power series. We focus on the bracket,

$$(\mathbf{I} + \Delta\mathbf{L}\mathbf{\Gamma})^{-1} \approx \mathbf{I} - \Delta\mathbf{L}\mathbf{\Gamma} + \Delta\mathbf{L}\mathbf{\Gamma}\Delta\mathbf{L}\mathbf{\Gamma} - \Delta\mathbf{L}\mathbf{\Gamma}\Delta\mathbf{L}\mathbf{\Gamma}\Delta\mathbf{L}\mathbf{\Gamma}\cdots \tag{12.114}$$

Thus,

$$\mathbf{L}^* = \langle\mathbf{L}\rangle - \langle\Delta\mathbf{L}\mathbf{\Gamma}\Delta\mathbf{L}\rangle + \langle\Delta\mathbf{L}\mathbf{\Gamma}\Delta\mathbf{L}\mathbf{\Gamma}\Delta\mathbf{L}\rangle - \langle\Delta\mathbf{L}\mathbf{\Gamma}\Delta\mathbf{L}\mathbf{\Gamma}\Delta\mathbf{L}\mathbf{\Gamma}\Delta\mathbf{L}\rangle\cdots \tag{12.115}$$

is a possible series expansion.

12.6.2 Example: Laminate

The following numerical example uses equations (12.108) and (12.115) as if they involve only matrix (tensor) products instead of operators. This works only for special microstructures. Here, we consider a laminate, for which all wave vectors with nonzero coefficients are parallel to the laminate normal. Therefore, only one $\mathbf{K}$ tensor appears that can effectively be factored out due to its idempotency.

We consider the conductivity of a laminate made of isotropic phases. The conductivities correspond to the indices,

$$\mathbf{L}_1 = \mathbf{I} \tag{12.116}$$

$$\mathbf{L}_2 = 2\mathbf{I}, \tag{12.117}$$

the phase distribution is

$$\chi_1(x) = \mathrm{H}(\mathrm{mod}(x, 2\pi) - \pi). \tag{12.118}$$

$\mathrm{H}(\cdot)$ is the Heaviside[9]-function or the unit step, which is 0 for negative and 1 for positive inputs. For the argument zero, different conventions exist, but we do not need this here. The indicator function is a square wave $\chi_1(x_1)$ as drawn in Figure 12.4. We chose $\mathbf{L}_0 = \mathbf{L}_2$. Then $\Delta\mathbf{L}|_2 = \mathbf{O}$ in phase 2 and $\Delta\mathbf{L}|_1 = \mathbf{L}_1 - \mathbf{L}_2$ in phase 1. Hence, $\Delta\mathbf{L}(\mathbf{x}) = \chi_1(\mathbf{x})(\mathbf{L}_1 - \mathbf{L}_2)$. The volume fractions are $v_1 = v_2 = 1/2$.

Since we have the $\boldsymbol{\Gamma}$-operators only in Fourier space, we need to express $\Delta\mathbf{L}$ in Fourier space as well. It is clear that the square wave in the direction of x_1 contains only wave vectors parallel to $\mathbf{e}_1$. The Fourier series of the square wave is

$$\chi_1(x_1) = \frac{1}{2} - \frac{2}{\pi}\left(\sin(x_1) + \frac{\sin(3x_1)}{3} + \frac{\sin(5x_1)}{5}\ldots\right), \tag{12.119}$$

or

$$\chi_1(x_1) = \frac{1}{2} - \frac{\mathrm{i}e^{-\mathrm{i}x_1}}{\pi} + \frac{\mathrm{i}e^{\mathrm{i}x_1}}{\pi} - \frac{\mathrm{i}e^{-3\mathrm{i}x_1}}{3\pi} + \frac{\mathrm{i}e^{3\mathrm{i}x_1}}{3\pi} - \frac{\mathrm{i}e^{-5\mathrm{i}x_1}}{5\pi} + \frac{\mathrm{i}e^{5\mathrm{i}x_1}}{5\pi}\cdots \tag{12.120}$$

in exponential form; see Figure 12.4. The form is quite simple since we have chosen volume fractions of 1/2. With this, the wave vectors $\mathbf{k}$ of the indicator function $\hat{\chi}_{1\mathbf{k}}$ in Fourier space are identified,

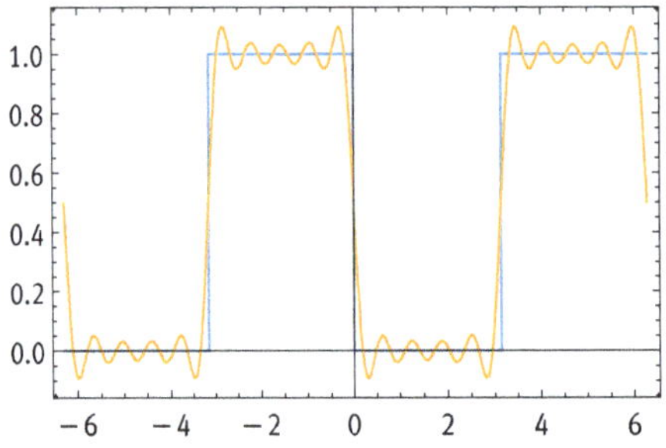

Figure 12.4: Indicator function according to equation (12.118) as a Fourier series up to order 10.

9 Oliver Heaviside, 1850–1925 (https://en.wikipedia.org/wiki/Oliver_Heaviside).

$$\mathbf{k} = \{0, \pm1, \pm3, \pm5\ldots\}\mathbf{e}_1 \tag{12.121}$$

$$\hat{\chi}_{1k} = \{1/2, \pm \mathrm{i}(\pi)^{-1}, \pm \mathrm{i}(3\pi)^{-1}, \pm \mathrm{i}(5\pi)^{-1}\ldots\}. \tag{12.122}$$

One notes that for $\mathbf{k} = \mathbf{o}$ the Fourier coefficient $\hat{\chi}_{1,0} = 1/2$ is the homogeneous part of $\chi_1(x_1)$, i. e., the volume fraction v_1 of phase 1. This is multiplied with $\mathbf{\Gamma}$, which results in a removal of the divergence-free part and the homogeneous part,

$$\mathbf{\Gamma} = \mathbf{\Gamma}_{\times\nabla}\mathbf{L}_0^{-1} \tag{12.123}$$

$$= v_2(2\mathbf{k}\cdot\mathbf{k})^{-1}\mathbf{k}\otimes\mathbf{k} \tag{12.124}$$

$$= \frac{1}{4}\mathbf{e}_1\otimes\mathbf{e}_1. \tag{12.125}$$

The $\mathbf{\Gamma}_0$-operator preserves the homogeneous part v_1 and removes the fluctuating part v_2,

$$\mathbf{\Gamma}_0 = v_1\mathbf{I}. \tag{12.126}$$

The expressions are simple because we have only one normalized wave vector. Generally, one needs to integrate or sum over all $\mathbf{K}$. This integration is usually decomposed into a path along rays from the origin to infinity and angles that parameterize the unit sphere. Here, the integration over the unit sphere vanishes since only one direction $\mathbf{k}/|\mathbf{k}| = \pm\mathbf{e}_1$ has nonzero Fourier coefficients. We merely need to sum up the coefficients in equation (12.122) without the homogeneous part $k = 0$ according to the definition of the scalar product in Fourier space,

$$\sum_{k=-\infty\cdots\infty, k\neq 0} \hat{\chi}_{1k}\hat{\chi}_{1k}^{\dagger} = 2\left(\frac{1}{\pi^2} + \frac{1}{3^2\pi^2} + \frac{1}{5^2\pi^2}\cdots\right) = \frac{1}{4}, \tag{12.127}$$

where the factor two appears due to the equality of the coefficients for k and $-k$. This is simply the product v_1v_2. The factor v_2 stems from the fact that we consider $\hat{\chi}_1$, the factor v_1 is due to the removal of the homogeneous part.

The exact solution

After preparing all the expressions, we are ready to plug these into equation (12.108):

$$\mathbf{L}^* = \mathbf{L}_0 + \mathbf{\Gamma}_0(\mathbf{I} + \Delta\mathbf{L}\mathbf{\Gamma})^{-1}\Delta\mathbf{L} \tag{12.128}$$

$$= 2\mathbf{I} + \frac{1}{2}\mathbf{I}\left(\mathbf{I} + (-\mathbf{I})\frac{1}{4}\mathbf{e}_1\otimes\mathbf{e}_1\right)^{-1}(-\mathbf{I}) \tag{12.129}$$

$$= 2\mathbf{I} - \frac{1}{2}\mathbf{I}\left(\mathbf{I} - \frac{1}{4}\mathbf{e}_1\otimes\mathbf{e}_1\right)^{-1} \tag{12.130}$$

$$= 2\mathbf{I} - \frac{1}{2}\mathbf{I}\left(\mathbf{I} + \frac{1}{3}\mathbf{e}_1\otimes\mathbf{e}_1\right) \tag{12.131}$$

$$= \begin{bmatrix} 2-\frac{2}{3} & & \\ & 2-\frac{1}{2} & \\ & & 2-\frac{1}{2} \end{bmatrix} = \operatorname{diag}(4/3, 3/2, 3/2). \tag{12.132}$$

The results correspond to the well-known analytic solution, where in the direction of lamination x_1, a serial composition of the conductivities is obtained $(v_1 L_1^{-1} + v_2 L_2^{-1})^{-1} = 4/3$ and perpendicular to the lamination direction a parallel composition $v_1 L_1 + v_2 L_2 = 3/2$ is obtained. These calculations are performed by the script given in Listing 12.3.

Listing 12.3: Mathematica notebook (https://gitlab.com/gluegerainer/listings-homogenization-methods/-/blob/main/Listing-17_en.nb) for the numeric example of the laminate conductivity, part 1.

```
Remove["Global`*"]
(* parameters *)
l1 = 1; (* conductivities *)
l2 = 2;
L1 = l1 IdentityMatrix[3];
L2 = l2 IdentityMatrix[3];
DL = L1 - L2;
v1 = 1/2;  (* volume fractions *)
v2 = 1 - v1;

(* indicator function for phase 1 *)
ID = UnitStep[Mod[x1, 2 Pi] - Pi 2 v2];

(* The Fourier series is not needed but given for clarity *)
ord = 10;  (* coefficients up to +- 10 *)
koeffs =  Table[FourierCoefficient[ID, x1, i], {i, -ord, ord}]
exp = Plot[{ID, Sum[Exp[I i x1] koeffs[[i + ord + 1]], {i, -ord, ord}]},
           {x1, -2 Pi, 2 Pi}]

Print["The 0-Fourier coefficient corresponds to the volume fraction v_1"];
Print["koeffs[[ord+1]]-v1=", koeffs[[ord + 1]] - v1];
koeffs = Delete[koeffs, ord + 1]; (* remove homogeneous part, i.e.\ the 0-coefficient *)
Print["The sum over the remaining coefficients approaches the product v_1 v_2"];
Print["koeffs.Conjugate[koeffs]-v1 v2//N=", koeffs.Conjugate[koeffs] - v1 v2 // N];

(* the only relevant normalized wave vector *)
k = {1, 0, 0};
GAMMA = v2 Outer[Times, k, k]/l2; (* Divide by the reference conductivity l0=l2=2*)

(* exact solution with L0=L2 as the reference conductivity *)
L2 + v1 Inverse[IdentityMatrix[3] + DL.GAMMA].DL
```

The series expansion

The series expansion is according to equation (12.115),

$$\mathbf{L}^* = \mathbf{L}_1 - \mathbf{L}_2 + \mathbf{L}_3 - \mathbf{L}_4 \cdots \tag{12.133}$$

with the expressions:

$$\mathbf{L}_1 = \langle \mathbf{L} \rangle = v_1 \mathbf{L}_1 + v_2 \mathbf{L}_2 = \operatorname{diag}(3/2, 3/2, 3/2) \tag{12.134}$$

$$\mathbf{L}_2 = v_1 \boldsymbol{\Gamma} \Delta \mathbf{L} \boldsymbol{\Gamma} = \operatorname{diag}(1/8, 0, 0) \tag{12.135}$$

$$\mathbf{L}_3 = v_1 \mathbf{\Gamma}\Delta\mathbf{L}\mathbf{\Gamma}\Delta\mathbf{L}\mathbf{\Gamma} = \mathrm{diag}(-1/32, 0, 0) \tag{12.136}$$

$$\mathbf{L}_4 = v_1 \mathbf{\Gamma}\Delta\mathbf{L}\mathbf{\Gamma}\Delta\mathbf{L}\mathbf{\Gamma}\Delta\mathbf{L}\mathbf{\Gamma} = \mathrm{diag}(1/128, 0, 0) \tag{12.137}$$

$$\vdots$$

One can check that the sum of all $\mathbf{L}_i$ approaches the exact solution. The complement to Listing 12.3 is given in Listing 12.4.

Listing 12.4: Mathematica notebook (https://gitlab.com/gluegerainer/listings-homogenization-methods/-/blob/main/Listing-17_en.nb) for the numeric example of the laminate conductivity, part 2.

```
(* Series expansion *)
Lseries1 = v1  L1 + v2 L2
Lseries2 = v1 DL.GAMMA.DL
Lseries3 = v1 DL.GAMMA.DL.GAMMA.DL
Lseries4 = v1 DL.GAMMA.DL.GAMMA.DL.GAMMA.DL
Lseries5 = v1 DL.GAMMA.DL.GAMMA.DL.GAMMA.DL.GAMMA.DL
Lseries6 = v1 DL.GAMMA.DL.GAMMA.DL.GAMMA.DL.GAMMA.DL.GAMMA.DL
Lseries7 = v1 DL.GAMMA.DL.GAMMA.DL.GAMMA.DL.GAMMA.DL.GAMMA.DL.GAMMA.DL
Print["========"]
Lseries1 // N
Lseries1 - Lseries2 // N
Lseries1 - Lseries2 + Lseries3 // N
Lseries1 - Lseries2 + Lseries3 - Lseries4 // N
Lseries1 - Lseries2 + Lseries3 - Lseries4 + Lseries5 // N
Lseries1 - Lseries2 + Lseries3 - Lseries4 + Lseries5 - Lseries6 // N
Lseries1 - Lseries2 + Lseries3 - Lseries4 + Lseries5 - Lseries6 + Lseries7 // N
```

It gives the following output:

```
{{3/2,0,0},{0,3/2,0},{0,0,3/2}}
{{1/8,0,0},{0,0,0},{0,0,0}}
{{-(1/32),0,0},{0,0,0},{0,0,0}}
{{1/128,0,0},{0,0,0},{0,0,0}}
{{-(1/512),0,0},{0,0,0},{0,0,0}}
{{1/2048,0,0},{0,0,0},{0,0,0}}
{{-(1/8192),0,0},{0,0,0},{0,0,0}}
========
{{1.5,0.,0.},{0.,1.5,0.},{0.,0.,1.5}}
{{1.375,0.,0.},{0.,1.5,0.},{0.,0.,1.5}}
{{1.34375,0.,0.},{0.,1.5,0.},{0.,0.,1.5}}
{{1.3359375,0.,0.},{0.,1.5,0.},{0.,0.,1.5}}
{{1.333984375,0.,0.},{0.,1.5,0.},{0.,0.,1.5}}
{{1.33349609375,0.,0.},{0.,1.5,0.},{0.,0.,1.5}}
{{1.33337402344,0.,0.},{0.,1.5,0.},{0.,0.,1.5}}
```

12.6.3 Discrete microstructures

The advantage of the Fourier method is that for microstructural data that lies on a Cartesian grid, fast numerical discrete transformations are available. For example, in Mathematica, we can call the function `Fourier` on a 3D grid with data points, which produces as many Fourier coefficients as data points. Similar functions are available in most other computer-aided mathematics tools.

As an example, we consider a periodic microstructure without rotational symmetry as in Figure 12.5. The conductivities of the phases are

$$\mathbf{L}_1 = \begin{bmatrix} 3 & 1 & 2 \\ 1 & 3 & 1 \\ 2 & 1 & 3 \end{bmatrix} \mathbf{e}_i \otimes \mathbf{e}_j, \quad \mathbf{L}_2 = \begin{bmatrix} 2 & 0 & 1 \\ 0 & 2 & 1 \\ 1 & 1 & 2 \end{bmatrix} \mathbf{e}_i \otimes \mathbf{e}_j. \tag{12.138}$$

Both conductivities are orthotropic and not aligned with the periodicity frame, which is the most general case that can be considered in this setting.

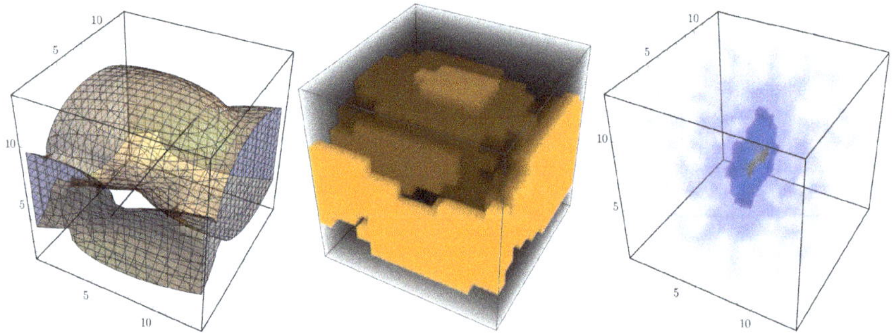

Figure 12.5: A unit cell of the microstructure. Left: analytic, Center: rasterized, Right: Magnitude of the Fourier coefficients.

The reference conductivity is chosen as $\mathbf{L}_0 = l_0\mathbf{I}$ with $l_0 = 3$. This value is close to the conductivities $\mathbf{L}_{1,2}$. In case of series expansions or iteration methods (see Section 12.7.2), the convergence rate depends on the choice of $\mathbf{L}_0$. The analytic result is insensitive w. r. t. this choice. However, it is nevertheless reasonable to use a reference conductivity close to the phases' conductivities to have a numerically well-conditioned linear system. The Mathematica notebook in Listing 12.5 generates the Fourier coefficients for the field $\Delta\mathbf{L}$.

Listing 12.5: Mathematica notebook (https://gitlab.com/gluegerainer/listings-homogenization-methods/-/blob/main/Listing-19_en.nb) for calculating the tensorial Fourier coefficients for a two-phase microstructure.

```
Remove["Global`*"]
(* local conductivities and reference conductivity *)
L1 = {{3, 1, 2}, {1, 3, 1}, {2, 1, 3}};
```

```
L2 = {{2, 0, 1}, {0, 2, 1}, {1, 1, 2}};
l0 = 3;
L0 = l0 IdentityMatrix[3];

(* fix a discretization, works for any size but odd sizes are easier to draw *)
size = 13;

(* Indicator function ID of the microstructure *)
radius = 0.4;
center = (size + 1)/2;
unitcell[x_] := Boole[((x[[1]] + 3 x[[2]])/4 - center)^2 +
                      ((x[[2]] + 3 x[[3]] - 2 x[[1]])/4 - center)^2 +
                      (x[[3]] - center)^2 <= (radius size)^2];
shifts = Tuples[{-size, 0, size}, 3]; (* periodic shifts *)
(* unit cell and neighboring cells *)
ID[x_] := Max[Table[unitcell[x + shifts[[i]]], {i, 1, 27}]]

(* graphic of the microstructure *)
image = RegionPlot3D[
   ID[{x, y, z}] > 0, {x, 1 - 0.5, size + 0.5}, {y, 1 - 0.5,
   size + 0.5}, {z, 1 - 0.5, size + 0.5}]

(* rasterization of the microstructure and graphic *)
ind = Table[ID[{i, j, k}], {k, 1, size}, {j, 1, size}, {i, 1, size}];
image = Image3D[ind, Boxed -> True]

(* Fourier coefficients of the conductivity *)
data = Table[ID[{i, j, k}] (L2 - L1) + L1 - L0, {i, 1, size}, {j, 1, size}, {k, 1, size}];
fdata = Table[Fourier[data[[;; , ;; , ;; , i, j]],
             FourierParameters -> {1, 1}]/size^3, {i, 1, 3}, {j, 1, 3}];

(* offset correction for the graphic *)
RL[arg_] := RotateLeft[arg, center]; (*works only for odd sizes*)
fdataOK = Table[RL[Map[RL, fdata[[i, j, ;; , ;; , ;;]], 2]], {i, 1, 3}, {j, 1, 3}];
ListDensityPlot3D[Abs[fdataOK[[1, 1, ;; , ;; , ;;]]]]
```

The reconstruction of the indicator function from the Fourier data shows that the discrete Fourier transform reproduces the exact function values at the grid points. In Figure 12.6, ΔL_{11} is plotted parallel to an edge of the unit cell along the line in Figure 12.5 left. This is not needed below but anyway good to test.

Listing 12.6: Reconstruction (https://gitlab.com/gluegerainer/listings-homogenization-methods/-/blob/main/Listing-19_en.nb) of the rasterized microstructure from the discrete Fourier data (see Figure 12.6, continuation of Listing 12.5).

```
f[x1_, x2_, x3_] = Sum[Exp[-I {i-1, j-1, k-1}.{x1-1, x2-1, x3-1} 2 Pi/size] fdata[[1, 1, i, j, k
    ]],
              {i, 1, size}, {j, 1, size}, {k, 1, size}];
p = Show[Plot[Re[f[center, x2, center]], {x2, 0, size},
         PlotRange -> All, Frame -> True], (*works only for odd sizes*)
      Graphics[Join[{Red, PointSize[Large]},
              Table[Point[{i, data[[center, i, center, 1, 1]]}], {i, 1, size}]]]]
```

Next, we set up the linear system according to equation (12.113); see Listing 12.7. This is the core of the method. Indices outside the discretization d are taken modulo d according to the periodic continuation in real- and Fourier space.

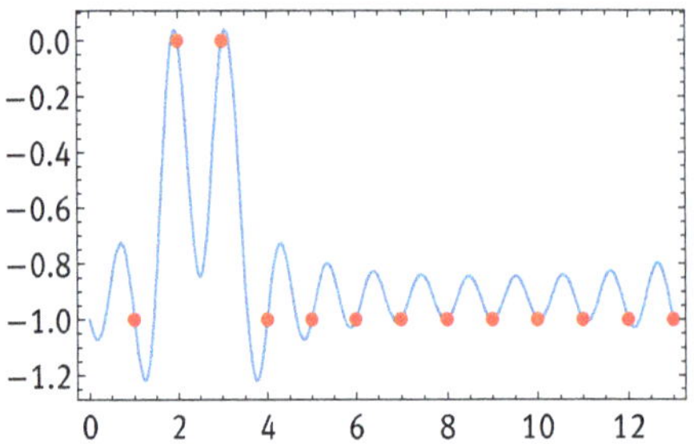

Figure 12.6: Reconstruction of the rasterized microstructure from the discrete Fourier data along a line through the unit cell.

Listing 12.7: Construction (https://gitlab.com/gluegerainer/listings-homogenization-methods/-/blob/main/Listing-19_en.nb) of the linear system for the discrete Fourier method, continuation of Listing 12.5.

```
(* initialize variables *)
vars = Table[Lstar[Min[{m, n}], Max[{m, n}], i, j, k],
            {i, 1, size}, {j, 1, size}, {k, 1, size},
            {m, 1, 3}, {n, 1, 3}];
(* initialize projector K as a function of k *)
K[k_] := Outer[Times, k, k]/k.k;
(* prepare all possible c- and k-vectors except k={0,0,0} *)
clist = Flatten[    Table[{i, j, k}, {i, 1, size}, {j, 1, size}, {k, 1, size}], 2];
klist = DeleteCases[clist, {1, 1, 1}];
(* generate tensorial equation, Mod due to periodicity in Fourier space *)
eqs = {};
Table[
    AppendTo[eqs, vars[[c[[1]], c[[2]], c[[3]]]]+
        Sum[fdata[[;; , ;; ,
                   Mod[c[[1]] - k[[1]], size] + 1,
                   Mod[c[[2]] - k[[2]], size] + 1,
                   Mod[c[[3]] - k[[3]], size] + 1]].
            K[k - {1, 1, 1}].vars[[k[[1]], k[[2]], k[[3]]]]/l0, {k, klist}]
        -fdata[[;; , ;; , c[[1]], c[[2]], c[[3]]]]],
    {c, clist}];
(* we should have size^3 tensorial equations *)
{Length[eqs], size^3}
(* we should have 6 * size^3 scalar equations and variables *)
eqlst = Flatten[Table[eqs[[i, m, n]] == 0, {i, 1, Length[eqs]}, {m, 1, 3}, {n, m, 3}]];
6 size^3
Length[eqlst]
varlst = DeleteDuplicates[Flatten[vars]];
Length[varlst]
```

The solution of this linear system is obtained with Mathematica-internal methods, see Listing 12.8.

Listing 12.8: Solution (https://gitlab.com/gluegerainer/listings-homogenization-methods/-/blob/main/Listing-19_en.nb) of the linear system, continuation of Listing 12.7.

```
erg = Solve[eqlst, varlst];
LEFF = L0 + {
    {erg[[1, 1, 2]], erg[[1, 2, 2]], erg[[1, 3, 2]]},
    {erg[[1, 2, 2]], erg[[1, 4, 2]], erg[[1, 5, 2]]},
    {erg[[1, 3, 2]], erg[[1, 5, 2]], erg[[1, 6, 2]]}
```

```
  };
MatrixForm[LEFF // Chop]
```

12.6.4 A linear solver for an individual component

Our problem has a special feature: We are interested only in a single solution component, namely the homogeneous part of the solution due to the averaging in equation (12.108), i. e., we are only interested in the coefficients to $\mathbf{k} = \mathbf{o}$ of the $\mathbf{L}_\star$-field. These are only 6 out of the $6 \times d^3$ variables. Standard solvers determine the entire solution vector, which is unnecessarily expensive from a computational point of view. Therefore, in this section, we demonstrate a solution method by Lee et al. (2014), which is a Jacobi iteration for a single solution component. We write the linear system as

$$\mathbf{M} \cdot \mathbf{x} = \mathbf{f}, \tag{12.139}$$

with the matrix $\mathbf{M}$, the unknown solution $\mathbf{x}$ and the right-hand side $\mathbf{f}$. Necessary for convergence is the positive definiteness of the coefficient matrix. Sufficient for convergence is strong diagonal dominance. The method may nevertheless converge even without diagonal dominance. We perform a Jacobi preconditioning and define the iteration matrix $\mathbf{G}$

$$\mathbf{J} = \operatorname{diag}(\mathbf{M}) \tag{12.140}$$

$$\mathbf{G} = -\mathbf{J}^{-1}(\mathbf{M} - \mathbf{J}) \tag{12.141}$$

which is part of a Neumann-series expansion of $\mathbf{x}$. The iteration is executed by applying $\mathbf{G}$ to the initial vector $\mathbf{r}_0$, which is entirely filled with zeros except for the index k of the sought solution, where $r_k = 1$. This is repeated until $\mathbf{r}_n$ is sufficiently close to the zero vector,

$$\mathbf{r}_i = \mathbf{r}_{i-1}\mathbf{G}, \tag{12.142}$$

$$\mathbf{r}_n = \mathbf{r}_0\mathbf{G}^n. \tag{12.143}$$

Because of $\mathbf{G}^n$, convergence can be guaranteed if the largest eigenvalue (spectral radius) of $\mathbf{G}$ is smaller than one. The solution component is then

$$x_k = \left(\sum_{i=0}^{n} \mathbf{r}_i\right) \cdot \mathbf{J}^{-1} \cdot \mathbf{f}. \tag{12.144}$$

It is easy to see that this is more efficient than a direct solution for the entire solution vector. An LU-decomposition requires approximately $2N^3/3$ operations (additions and multiplications) for fully occupied linear systems of size N. A single iteration requires $2N^2$ operations. Since we need only six solution components and much less than N iterations, the iterative solution method is much more efficient. It is implemented in Listing 12.9.

Listing 12.9: Iterative solution (https://gitlab.com/gluegerainer/listings-homogenization-methods/-/blob/main/Listing-19_en.nb) of a single solution component of $\mathbf{L}_\star$ of the linear system of the Fourier method, continuation of Listing 12.7.

```
(* extraction of the coefficient matrix and the right-hand side *)
ms = CoefficientArrays[eqlst, varlst];
(* matrix plot of the system matrix *)
plot = MatrixPlot[ms[[2]], PlotLegends -> True]
(* prepare iteration *)
Jaci = SparseArray[DiagonalMatrix[Table[ms[[2, i, i]], {i, 1, Length[ms[[1]]]}]]];
InvJaci = SparseArray[DiagonalMatrix[1/Table[ms[[2, i, i]], {i, 1, Length[ms[[1]]]}]]];
G = -InvJaci.(ms[[2]] - Jaci);
z = - InvJaci.ms[[1]] (* The minus is needed because CoefficientArrays
                         assumes the normal form  r + M.x == 0 *)
(* collect the components of the effective conductivity in Leff *)
Leff = Table[0, 3, 3];
(* Solve for the 6 solution components *)
For[ii = 1, ii <= 3, ii++,
    For[jj = ii, jj <= 3, jj++,
        (* find index *)
        index = Position[varlst, Lstar[ii, jj, 1, 1, 1]][[1,1]];
        (* initialize starting vector *)
        p = SparseArray[Table[0, {i, 1, Length[ms[[1]]]}]];
        r = p;
        r[[index]] = 1;
        (* iterate *)
        While[Norm[r] > 0.000001,
              p = p + r;
              r = r.G];
        result = p.z;
        Leff[[ii, jj]] = Chop[L0[[ii, jj]] + result];
        Leff[[jj, ii]] = Leff[[ii, jj]];
        ]]
Leff//MatrixForm
```

12.7 Spectral solver: Fixed-point iteration for RVE problems

The method presented in the preceding section has the disadvantage that, as an intermediate step, a large, nonsymmetric, fully populated linear system with complex coefficients needs to be solved. The size of the system is $k \times d^3$, with the edge discretization d and the number of coefficients in the constitutive tensor to be homogenized, e. g., $k = 6$ for heat conduction and $k = 21$ for linear elasticity. Considering real and imaginary parts individually, the coefficient matrix contains $2(kn^3)^2$ floating point numbers. Even for small n, the memory requirement becomes unacceptable quickly.

Therefore, we now consider an iterative method that is usually referred to as the Fourier method or spectral method. It does not require the setup of a humongous linear system. Instead, it solves specific boundary value problems iteratively, from which the effective coefficients of the sought constitutive tensor are extracted. We will be able to consider much finer discretizations d.

12.7.1 Fixed-point iteration for solving a differential equation: A 1D example

Before considering the 3D case, let us examine a 1D example where we are interested in the effective conductivity of a layered composite. The heat flux $q(x)$ is proportional to the temperature gradient $T'(x)$ with the factor of proportionality $l(x)$ being the conductivity but in the opposite direction of the temperature gradient,

$$q(x) = -l(x)T'(x). \tag{12.145}$$

In the stationary case, it is

$$q'(x) = 0. \tag{12.146}$$

Together this becomes

$$0 = \left(l(x)T'(x)\right)' \tag{12.147}$$

$$= l'(x)T'(x) + l(x)T''(x). \tag{12.148}$$

The boundary conditions are the temperatures $T(x_0) = T_0$ and $T(x_1) = T_1$ at the points $x_0 < x_1$.

The analytic solution

We substitute $T'(x) = P(x)$ and have by this an ordinary linear first-order differential equation,

$$P'(x) = [-l'(x)/l(x)]P(x). \tag{12.149}$$

Its solution is

$$P(x) = P_0 \exp\left(\int -l'(x)/l(x)\,\mathrm{d}x\right), \tag{12.150}$$

which can be simplified due to the special integrand,

$$P(x) = P_0 \exp(-\ln(l(x))) = \frac{P_0}{l(x)}. \tag{12.151}$$

The heat flux is

$$q(x) = -l(x)T'(x) = -P_0 = \bar{q}. \tag{12.152}$$

It does not depend on x, otherwise $q'(x) = 0$ would not be satisfied. This is a special property of the 1D case. Integrating $P(x)$ once gives $T(x)$ and the constant of integration T_0:

$$T(x) = \int P(x)\,\mathrm{d}x = P_0 \int_{x_0}^{x} 1/l(\underline{x})\,\mathrm{d}\underline{x} + T_0. \tag{12.153}$$

The latter is adapted together with P_0 to the boundary conditions. Subtracting T_0 and inserting $x = x_1$ one obtains the temperature difference $\Delta T = T_1 - T_0$ on the left-hand side. On the right-hand side, one can expand with $\Delta x/\Delta x$, where $\Delta x = x_1 - x_0$. We replace $P_0 = -\overline{q}$,

$$\Delta T = -\overline{q}\frac{\Delta x}{\Delta x} \int_{x_0}^{x_1} 1/l(\underline{x})\,\mathrm{d}\underline{x}. \tag{12.154}$$

We can finally summarize $\Delta T/\Delta x$ to $\overline{T'}$,

$$-\overline{q} = \overline{T'} \underbrace{\left(\frac{1}{\Delta x} \int_{x_0}^{x_1} l(\underline{x})^{-1}\,\mathrm{d}\underline{x} \right)^{-1}}_{l^*}, \tag{12.155}$$

where we can identify the effective conductivity l^* as the Reuss mean of the local conductivity $l(x)$. This is the expected result since, in the 1D case, only a serial composition of the phases is possible.

Fixed-point iteration

How could one obtain an approximate solution for $T(x)$ without solving the differential equation analytically? We can solve equation (12.148) for $T''(x)$ and integrate twice to obtain $T(x)$. Repeating this is a fixed-point iteration,

$$T_{n+1}(x) = -\int\int l'(x)T_n'(x)/l(x)\,\mathrm{d}x\,\mathrm{d}x. \tag{12.156}$$

After each integration, the constants of integration need to be adapted to the boundary conditions. Formally, with each iteration, the expressions $T_2(x), T_3(x)\ldots$ become more complex, but a computer algebra system can handle this. In Listing 12.10, an implementation is given for a linearly growing conductivity $l(x) = K_0 + K_1 x$ and unit intervals $x = 0\ldots 1$, $T(0) = 0$, $T(1) = 1$. We use $T_1(x) = x$ as the starting solution, which satisfies the boundary conditions. Physical units are dropped for simplicity. The result is plotted in Figure 12.7. One observes convergence toward the solution $T(x) = \frac{\ln(K_0+K_1x)}{\ln(K_0+K_1)}$, by which we have $l^* = K_1/\ln(K_0 + K_1)$. For $K_1 = 30$ and $K_0 = 1$, we have $l^* \approx 8.7362$. The effective conductivity of the iterated solution $T_n'(x)l(x)$ can be probed at an arbitrary point in the interval $x = 0\ldots 1$.

Listing 12.10: Mathematica notebook (https://gitlab.com/gluegerainer/listings-homogenization-methods/-/blob/main/Listing-24_en.nb) for the fixed-point iteration of a simple 1D boundary value problem.

```
K = 30; T0 = 0; T1 = 1;
l[x_] = K x + 1;
refsol = DSolve[{D[l[x] T'[x], x] == 0, T[0] == 0, T[1] == 1}, T[x], x][[1, 1, 2]] //
    FullSimplify
D[refsol, x] l[x] // N
current = x;  (* Initial solution *)
plist = {current};
Do[ expr = Integrate[-D[current, x] D[l[x], x]/l[x], x, x] + c1 x + c2;
   current = (expr /. Solve[{(expr /. {x -> 0}) == 0, (expr /. {x -> 1}) == 1}, {c1, c2}][[1]])
     // N; (* fit c1, c2 to BC *)
   Print[D[current, x]*l[x] /. x -> 0.5]; (* output of approximate conductivity *)
   AppendTo[plist, current], 10]; (* save result, repeat 10 times *)
Plot[plist,{x,0,1}] (* show graphs *)
```

Adjusting the convergence by using a reference conductivity

The above method may converge too slowly or even diverge. Therefore, we now consider the difference problem, which contains a comparison or reference conductivity l_0. With the decomposition $l(x) = l_0 + \tilde{l}(x)$, the above differential equation (12.148) becomes

$$0 = \tilde{l}'(x)T'(x) + (l_0 + \tilde{l}(x))T''(x), \tag{12.157}$$

which can be solved for the $T''(x)$ that appears next to l_0. This gives the iterator

$$T_{n+1}(x) = -\int\int \frac{-\tilde{l}'(x)T_n'(x) - \tilde{l}(x)T_n''(x)}{l_0}\,\mathrm{d}x\,\mathrm{d}x. \tag{12.158}$$

One can see that the magnitude of the nonlinear part that appears in front of the linear part C_1x+C_2 is inversely proportional to l_0. For large values of l_0, the method converges slower, as exemplary depicted in Figure 12.7 on the right.

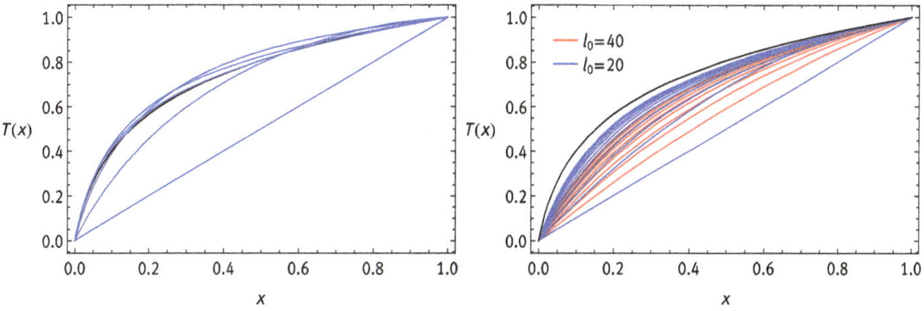

Figure 12.7: Convergence of the fixed-point iteration, left for equation (12.156) and right for equation (12.158), starting from $T_1(x) = x$ toward the solution $T(x) = \ln(1 + Kx)/\ln(1 + K)$ with $K = 30$. The second variant allows for adjusting the convergence rate via l_0 within limits.

12.7.2 Fixed-point iteration involving Fourier transforms

As has been discussed in Section 4.4, one usually assumes a periodically repeatable unit cell as the virtual material sample. The function space spanned by the Fourier base functions is optimally suited to represent fields on such an RVE. It is, for example, not necessary to enforce periodic boundary conditions. We will consider heat conduction on a periodic RVE. The starting point is

$$\mathbf{q} \cdot \nabla = 0, \quad \mathbf{q} = \mathbf{L} \cdot \mathbf{g}, \quad \mathbf{g} = \nabla T. \tag{12.159}$$

All fields are location-dependent. We evaluate the divergence of the heat flux with the decomposition $\mathbf{L} = \mathbf{L}_0 + \tilde{\mathbf{L}}$,

$$\mathbf{q} \cdot \nabla = (\mathbf{L}_0 \cdot \mathbf{g}) \cdot \nabla + \underbrace{(\tilde{\mathbf{L}} \cdot \mathbf{g})}_{\boldsymbol{\tau}} \cdot \nabla, \tag{12.160}$$

where the fluctuating part is written as $\boldsymbol{\tau}$ as in the difference problem. We next replace the functions by their Fourier representation,

$$\hat{\mathbf{q}} \cdot \nabla = (\mathbf{L}_0 \cdot \hat{\mathbf{g}}) \cdot \nabla + \hat{\boldsymbol{\tau}} \cdot \nabla. \tag{12.161}$$

In this representation, we can replace ∇ by $\mathrm{i}\mathbf{k}$,

$$\mathrm{i}\hat{\mathbf{q}} \cdot \mathbf{k} = \mathrm{i}(\mathbf{L}_0 \cdot \hat{\mathbf{g}}) \cdot \mathbf{k} + \mathrm{i}\hat{\boldsymbol{\tau}} \cdot \mathbf{k}. \tag{12.162}$$

Correspondingly, we have $\hat{\mathbf{g}} = \mathrm{i}\hat{T}\mathbf{k}$, which can be solved for $\hat{T}$,

$$\mathrm{i}\hat{\mathbf{q}} \cdot \mathbf{k} = -(\mathbf{L}_0 \cdot \hat{T}\mathbf{k}) \cdot \mathbf{k} + \mathrm{i}\hat{\boldsymbol{\tau}} \cdot \mathbf{k} \tag{12.163}$$

$$(\mathbf{L}_0 : \mathbf{k} \otimes \mathbf{k})\hat{T} = \mathrm{i}(\hat{\boldsymbol{\tau}} \cdot \mathbf{k} - \hat{\mathbf{q}} \cdot \mathbf{k}) \tag{12.164}$$

$$\hat{T} = \mathrm{i}(\hat{\boldsymbol{\tau}} \cdot \mathbf{k} - \hat{\mathbf{q}} \cdot \mathbf{k})/\kappa, \quad \kappa = \mathbf{L}_0 : \mathbf{k} \otimes \mathbf{k}. \tag{12.165}$$

Multiplying with $\mathrm{i}\mathbf{k}$ we obtain the temperature gradient on the left,

$$\hat{\mathbf{g}} = -(\hat{\boldsymbol{\tau}} \cdot \mathbf{k} - \hat{\mathbf{q}} \cdot \mathbf{k})\mathbf{k}/\kappa. \tag{12.166}$$

Only now we apply the equilibrium condition $\mathrm{i}\hat{\mathbf{q}} \cdot \mathbf{k} = 0$ in Fourier space ($\mathbf{q} \cdot \nabla = 0$ in real space). At the solution, we have

$$\hat{\mathbf{g}} = -\hat{\boldsymbol{\tau}} \cdot \mathbf{k} \otimes \mathbf{k}/\kappa. \tag{12.167}$$

This is used to eliminate $\hat{\boldsymbol{\tau}}$ in equation (12.166), where now $\hat{\mathbf{g}}$ appears twice in equation (12.166). We consider this as our fixed-point iterator after solving for one occurrence of $\hat{\mathbf{g}}$,

$$\hat{\mathbf{g}}_{n+1} = \hat{\mathbf{g}}_n + \hat{\mathbf{q}} \cdot \mathbf{k} \otimes \mathbf{k}/\kappa \tag{12.168}$$

$$\hat{\mathbf{g}}_n = \hat{\mathbf{g}}_{n+1} + \underbrace{\hat{\mathbf{q}} \cdot \mathbf{k} \otimes \mathbf{k}/\kappa}_{\Delta\hat{\mathbf{g}}} . \tag{12.169}$$

The update is $\hat{\mathbf{q}} \cdot \mathbf{k} \otimes \mathbf{k}/\kappa$. It is not evaluated for $\mathbf{k} = \mathbf{o}$, since the Fourier coefficient belongs to the homogeneous part $\overline{\mathbf{g}}$, which we assume as prescribed. To determine the update, one needs to insert the to be updated approximation $\mathbf{g}_n$ into the constitutive law $\mathbf{q}_n = \mathbf{L} \cdot \mathbf{g}_n$ in real space. Next, a Fourier transform is carried out, and the update term $\hat{\mathbf{q}}_n \cdot \mathbf{k} \otimes \mathbf{k}/\kappa$ is applied to obtain an improved approximation $\hat{\mathbf{g}}_{n+1}$. After a back transformation to real space, the next iteration starts. Constants of integration need not be adjusted to boundary conditions since both the function space and the boundary conditions are periodic.

This iteration requires a permanent switching between real space and Fourier space. In real space, the evaluation of the material law $\mathbf{q} = \mathbf{L} \cdot \mathbf{g}$ is local, and in Fourier space application of the nabla operator is local, hence the determination of $\Delta\hat{\mathbf{g}}$. Numerically, this is achieved effectively by the fast discrete Fourier transform. The fixed-point iteration taking advantage of the fast discrete Fourier transform has been proposed by Moulinec and Suquet (1998).

Example implementations of the above method are given in Listing 12.11 for Mathematica and in Listing 12.12 for Octave, which is largely compatible to MatLab. The same heat conduction problem as in Section 12.6.3 is considered. The relative size of the update $\|\Delta\mathbf{g}\|/\|\mathbf{g}\|$ converges quickly to zero (0.191, 0.016, 0.00266 . . .). The iteration converges slowly or even diverges for large phase contrasts.

The resulting heat flux field is plotted in Figure 12.8. One sees that the heat flux concentrates in the phase with the higher conductivity. From this solution, the homogeneous, average part $\overline{\mathbf{q}}$ is taken. The extraction of the effective tensorial conductivity $\mathbf{L}^*$ is done by imposing unit temperature gradients $\mathbf{g}$ in the three orthogonal directions $\mathbf{e}_i$ and solving three times for the resulting $\mathbf{q}$. By this, the effective tensorial conductivity can be identified columnwise in its implicit definition,

$$\overline{\mathbf{q}} = \mathbf{L}^* \cdot \overline{\mathbf{g}}. \tag{12.170}$$

Comments on the fixed-point iteration methods

Assignment of $\mathbf{g}_n$ and $\mathbf{g}_{n+1}$

It is clear that only one of the iterators (12.168, 12.169) can converge. This is typical for fixed-point iterations. Consider the example $x - x^2 = 0$, $x \geq 0$ with the solutions $x = 0$ and $x = 1$. There are two simple iterators, namely

$$x_{m+1} = x_m^2 \tag{12.171}$$

$$x_{m+1} = \sqrt{x_m}, \tag{12.172}$$

both have the fixed points 0 and 1. The first iterator converges for initial values $x_0 < 1$ to 0 and diverges for $x_0 > 1$, the second iterator converges to 1 for nonzero starting values.

In unclear cases, the converging iterator can be determined by testing, for example, by adjusting the sign of $\mathbf{L}_0$.

Versatility of the fixed-point iteration

In analysis, the fixed-point theorem is fundamental for deriving important theorems. However, the fixed-point iteration is often not considered a proper numerical method because of its poor convergence properties. It is also skimmed over quickly in seminars. This is unfortunate since it can be applied in very general spaces (complete metric spaces), including function spaces, as we have seen above.

Listing 12.11: Implementation (https://gitlab.com/gluegerainer/listings-homogenization-methods/-/blob/main/Listing-25_en.nb) of a fixed-point iteration for determining the effective conductivity in Mathematica.

```
Remove["Global`*"]
size = 16;
(* define microstructure *)
center = (size + 1)/2; radius = 0.4;
unitcell[x_]:=Boole[((x[[1]]+3x[[2]])/4-center)^2+((x[[2]]+3x[[3]]-2x[[1]])/4-center)^2+(x[[3]]-
     center)^2 <= (radius size)^2];
shifts = Tuples[{-size, 0, size}, 3];(* periodic shift *)
ID[x_] := Max[Table[unitcell[x + shifts[[i]]], {i, 1, 27}]]
(* define phases' conductivities *)
L1 = {{3, 1, 2}, {1, 3, 1}, {2, 1, 3}};
L2 = {{2, 0, 1}, {0, 2, 1}, {1, 1, 2}};
(* test case: laminate of isotropic phases
   ID[x_]:=Boole[x[[1]]<=size/2];L1=IdentityMatrix[3];L2=2L1; *)
L0 = (L1 + L2)/2;  (* fix L0 *)
gmean = {0.0, 0.0, 1.0};(* initialize the starting field *)
g = Table[gmean, size, size, size];
normdeltag = 1;
normg = 1;
While[(normdeltag/normg) > 0.005, (* stop when Norm[Delta g]/Norm[g]<0.005 *)
    q = Table[(L2 + ID[{i, j, k}] (L1 - L2)).g[[i, j, k]], {i, 1, size}, {j, 1, size}, {k, 1,
      size}]; (* calculate heat flux *)
    qdach = Table[Fourier[q[[;; , ;; , ;; , i]]], {i, 1, 3}]; (* fast Fourier transform *)
    deltagdach = Table[  (* calculate update without the 0,0,0-coefficient *)
    kk = {i - 1, j - 1, k - 1};
    If[Total[kk] > 0, kk qdach[[;; , i, j, k]].kk/(kk.L0.kk), {0,0,0}], {i, size}, {j, size}, {k
     , size}];
    deltagdach[[1, 1, 1]] = {0, 0, 0}; (* remove update of 0,0,0 coefficient *)
    deltag = Re[Table[InverseFourier[deltagdach[[;; , ;; , ;; , i]]], {i, 1, 3}]]; (* inverse
      fast FT *)
    normdeltag = Sum[Abs[deltag[[i, j, k, l]]], {i, 1, 3}, {j, 1, size}, {k, 1, size}, {l, 1,
      size}]; (* norm of the increment *)
    g = g - Table[deltag[[i, j, k, l]], {j, 1, size}, {k, 1, size}, {l, 1, size}, {i, 1, 3}]; (*
       update of the g-field *)
    normg = Sum[Abs[g[[j, k, l, i]]], {i, 1, 3}, {j, 1, size}, {k, 1, size}, {l, 1, size}]; (*
      norm of the g-field *)
    (* output for monitoring *)
    Print["============="];
    Print[(normdeltag/normg) // Chop];
    Print["Average g: ", Chop[Mean[Flatten[g, 2]]]];
    Print["Average q: ", Chop[Mean[Flatten[q, 2]]]];
    ]
```

Listing 12.12: Implementation (https://gitlab.com/gluegerainer/listings-homogenization-methods/-/blob/main/Listing-26_en.nb) of a fixed-point iteration for determining the effective conductivity in Octave (MatLab compatible).

```
# This is an implementation of a spectral solver. It solves the heat conduction DE
#
#   q(x).V = 0        (I)   V = nabla
#
# with
#
#   q(x)=L(x).g(x)    (II)
#
# with the average of g(x) prescribed.
# For each iteration q.V is calculated for (I) in Fourier space.
# (II) is evaluated in real space. The update of g(x) is
#
#   Delta_g = (q_hat.k) / (k.L0.k) k
#
# with L0 the reference conductivity l0 * identity matrix.
# L0 should not be too far from the volume average ||L(x)||.
# k is the wave vector.

clear;
clc;

page_output_immediately(1);

# size of the periodically repeated unit cell
size       = 20;

# initialize vector field
g          = zeros(3,size,size,size);
q          = zeros(3,size,size,size);
qdach      = zeros(3,size,size,size);
deltagdach = zeros(3,size,size,size);

# set the conductivities
l0    = 1.5;
L0    = eye(3,3).*l0;
L1    = 1;
L2    = 2;

# precision goal
tol   = 1e-8;

# The effective conductivities requires to prescribe the effective
# temperature gradient along three orthogonal directions.
# The field g is initialized homogeneously
for ii = 1:3
  gmean         = zeros(3,1);
  gmean(ii,1) = 1;
  for jj = 1:size
    for kk = 1:size
      for ll = 1:size
          g(1:3,jj,kk,ll)=gmean;
      end
    end
  end

```

```
# Start of the iteration
  residuum = 1;
  while residuum > tol
# determine local heat flux
    for jj = 1:size
      for kk = 1:size
        for ll = 1:size
# The if clause encodes the microstructure, i.e., the value of L(x),
# here as a laminate.
          if jj <= size / 2
            q(1:3,jj,kk,ll) = L1*g(1:3,jj,kk,ll);
          elseif jj > size /2
            q(1:3,jj,kk,ll) = L2*g(1:3,jj,kk,ll);
          end
        end
      end
    end

# Fourier transform of q
    qdach(1,:,:,:) = fftn(q(1,:,:,:))./sqrt(size*size*size);
    qdach(2,:,:,:) = fftn(q(2,:,:,:))./sqrt(size*size*size);
    qdach(3,:,:,:) = fftn(q(3,:,:,:))./sqrt(size*size*size);


# Calculation of the divergence by multiplication with k except for k = (0 0 0)
# which corresponds to the prescribed mean value.
    for jj = 1:size
      for kk = 1:size
        for ll = 1:size
          if (jj ~= 1 ) || (kk ~= 1) || (ll ~= 1)
            kkk = [ jj-1;
                    kk-1;
                    ll-1 ];
            deltagdach(:,jj,kk,ll) = ( kkk'*qdach(:,jj,kk,ll) )/( kkk'*L0*kkk )*kkk;
          else
            deltagdach(:,jj,kk,ll) = zeros(3,1);
          end
        end
      end
    end

# Fourier Backtransform
    deltag(1,:,:,:) = ifftn(deltagdach(1,:,:,:)).*sqrt(size*size*size);
    deltag(2,:,:,:) = ifftn(deltagdach(2,:,:,:)).*sqrt(size*size*size);
    deltag(3,:,:,:) = ifftn(deltagdach(3,:,:,:)).*sqrt(size*size*size);

# Update
    g=g-deltag;

# Calculate the proportional update sd/sg, where sg is scalar(g) and sd is scalar(delta g)
    sg=0;
    sd=0;
    for jj = 1:size
      for kk = 1:size
        for ll = 1:size
          sg = sg + sqrt( g(1,jj,kk,ll)^2 + g(2,jj,kk,ll)^2 + g(3,jj,kk,ll)^2 );
          sd = sd + sqrt( deltag(1,jj,kk,ll)^2 + deltag(2,jj,kk,ll)^2 + deltag(3,jj,kk,ll)^2 );
        end
      end
    end
```

```
    residuum=sd/sg
  end
  # The effective heat flux, which is the coefficient to k=(0 0 0), is extracted.
  qdach(:,1,1,1)./sqrt(size^3)
end
# The result is the effective laminate conductivity
#
#          [  1.333             ]
# L_eff  = [           1.5      ]
#          [                1.5 ]
#
# i.e. the arithmetic mean of (1,2) parallel to the laminate and the harmonic mean
# perpendicular to the laminate.
```

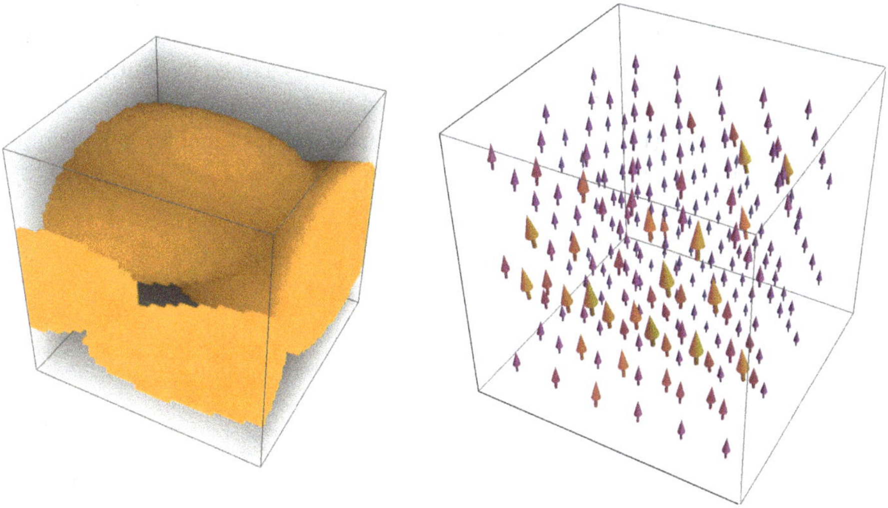

Figure 12.8: Left: Discretization of the microstructure for the spectral solver on a grid of size 40^3 with one phase hidden; see also Figure 12.5. Right: Heat flux field for $\mathbf{L}_1 = \mathbf{I}$, $\mathbf{L}_2 = 10\mathbf{L}_1$ and $\overline{\mathbf{g}} = \mathbf{e}_z$. One recognizes a heat flux concentration in the second phase.

12.8 Green's function method

Green's function method[10] for solving linear differential equations of the form $L(u(x)) = f(x)$ for $u(x)$ consists basically in writing the right side $f(x)$ as a superposition of Dirac distributions $f(x) = \int \delta(x - x')f(x')\,\mathrm{d}x'$. The solution is then the superposition of the solution $G(x, x')$ to the differential equation with the Dirac distribution $L(G(x, x')) = \delta(x - x')$ as the right side, which is called Green's function.

10 The same George Green postulated the existence of the elastic strain energy.

12.8.1 Green's function method example for the Laplace operator

We consider the isotropic homogeneous heat conduction problem with normalized conductivity

$$\nabla^2 T(\mathbf{x}) = f(\mathbf{x}). \tag{12.173}$$

The solution $G(\mathbf{x} - \mathbf{x}')$ of

$$\nabla^2 G(\mathbf{x} - \mathbf{x}') = \delta(\mathbf{x} - \mathbf{x}') \tag{12.174}$$

is called the fundamental solution or eigenfunction of the Laplace[11]-operator $\nabla^2 T(\mathbf{x}) = T_{,jj}(\mathbf{x})$. We ignore boundary conditions, hence the whole problem is translation invariant. Therefore, we can replace $\mathbf{r} = \mathbf{x} - \mathbf{x}'$. The solution is obtained by Fourier forward transform, algebraic solution in Fourier space and backward transform according to equations (12.28) and (12.29). We insert the unknown Green's function in its Fourier space representation into equation (12.174) and obtain

$$\nabla_{\mathbf{r}}^2 \frac{1}{(2\pi)^{3/2}} \int \mathrm{e}^{\mathrm{i}\mathbf{k}\cdot\mathbf{r}} \hat{G}(\mathbf{k})\, \mathrm{d}v_{\mathbf{k}} = \frac{1}{(2\pi)^{3/2}} \int \left[\frac{1}{(2\pi)^{3/2}} \right] \mathrm{e}^{\mathrm{i}\mathbf{k}\cdot\mathbf{r}}\, \mathrm{d}v_{\mathbf{k}}. \tag{12.175}$$

The content of the squared bracket is the Fourier transform of the Dirac distribution according the prefactor convention used here (equations 12.28 and 12.29). The application of the Laplace operator $\nabla_{\mathbf{r}}^2$ gives

$$\int [-\mathbf{k}\cdot\mathbf{k}\hat{G}(\mathbf{k})] \mathrm{e}^{\mathrm{i}\mathbf{k}\cdot\mathbf{r}}\, \mathrm{d}v_{\mathbf{k}} = \int \left[\frac{1}{(2\pi)^{3/2}} \right] \mathrm{e}^{\mathrm{i}\mathbf{k}\cdot\mathbf{r}}\, \mathrm{d}v_{\mathbf{k}}. \tag{12.176}$$

Since the Fourier base functions form an orthonormal function basis (see equation (12.7)), these integrals are equal if and only if the components in front of each base function are equal. A comparison of coefficients gives

$$\hat{G}(\mathbf{k}) = -(\mathbf{k}\cdot\mathbf{k}\,(2\pi)^{3/2})^{-1}. \tag{12.177}$$

The bigger problem is often the back transformation to real space $G(\mathbf{r})$. For this simple case, the back transformation can be given explicitly, making use of the isotropy of $\hat{G}(\mathbf{k})$ in $\mathbf{k}$,

$$G(\mathbf{r}) = \frac{-1}{(2\pi)^{3/2}} \int \frac{1}{(2\pi)^{3/2}} \frac{\mathrm{e}^{\mathrm{i}\mathbf{k}\cdot\mathbf{r}}}{\mathbf{k}\cdot\mathbf{k}}\, \mathrm{d}v_{\mathbf{k}}. \tag{12.178}$$

11 Pierre-Simon Laplace, 1749–1827 (https://en.wikipedia.org/wiki/Pierre-Simon_Laplace).

The easiest way is to write $\mathbf{k} = k_i \mathbf{e}_i$ in spherical coordinates, with $\mathbf{r} = r\mathbf{e}_3$ in direction of the north pole,

$$k_1 = k \sin\theta \cos\phi, \tag{12.179}$$

$$k_2 = k \sin\theta \sin\phi, \tag{12.180}$$

$$k_3 = k \cos\theta. \tag{12.181}$$

The scalar product $\mathbf{k} \cdot \mathbf{r}$ is

$$\mathbf{k} \cdot \mathbf{r} = kr \cos\theta. \tag{12.182}$$

Integration over the spherical coordinates requires inserting the Jacobian determinant

$$J = \det(\partial k_{1,2,3}/\partial(k, \theta, \phi)) = k^2 \sin\theta, \tag{12.183}$$

the origin of which is explained in detail in Section 13.5. We obtain

$$G(\mathbf{r}) = \frac{-1}{(2\pi)^3} \int_0^\infty \int_0^\pi \int_0^{2\pi} \frac{\mathrm{e}^{\mathrm{i}\mathbf{k}\cdot\mathbf{r}}}{\mathbf{k}\cdot\mathbf{k}} k^2 \sin\theta \, \mathrm{d}\phi \, \mathrm{d}\theta \, \mathrm{d}k \tag{12.184}$$

$$= \frac{-1}{(2\pi)^3} \int_0^\infty \int_0^\pi \int_0^{2\pi} \mathrm{e}^{ikr\cos\theta} \sin\theta \, \mathrm{d}\phi \, \mathrm{d}\theta \, \mathrm{d}k. \tag{12.185}$$

The inner integral over ϕ can be solved immediately. Next, we substitute $\cos\theta = s$, therefore, $\mathrm{d}s = -\sin\theta \, \mathrm{d}\theta$. The limits correspond to $\theta = 0 \to s = 1$ and $\theta = \pi \to s = -1$, which leads to

$$G(\mathbf{r}) = \frac{1}{(2\pi)^2} \int_0^\infty \int_1^{-1} \mathrm{e}^{ikrs} \, \mathrm{d}s \, \mathrm{d}k = \frac{-1}{(2\pi)^2} \int_0^\infty \left[\frac{\mathrm{e}^{ikrs}}{ikr} \right]_{-1}^{1} \mathrm{d}k \tag{12.186}$$

$$= \frac{-1}{(2\pi)^2} \int_0^\infty \frac{\mathrm{e}^{ikr} - \mathrm{e}^{-ikr}}{ikr} \, \mathrm{d}k. \tag{12.187}$$

With $\mathrm{e}^{iz} = \cos z + \mathrm{i} \sin z$ and the antisymmetry of the sine $\sin(-z) = -\sin z$, one can see that the numerator of the integrand is $\mathrm{e}^{ikr} - \mathrm{e}^{-ikr} = 2\mathrm{i}\sin(kr)$, where the factor $2\mathrm{i}$ can be canceled out,

$$G(\mathbf{r}) = \frac{-1}{2\pi^2} \int_0^\infty \frac{\sin(kr)}{kr} \, \mathrm{d}k. \tag{12.188}$$

A final substitution $t = kr$ with $\mathrm{d}t = r \, \mathrm{d}k$ results in

$$G(\mathbf{r}) = \frac{-1}{2\pi^2 r} \int_0^\infty \frac{\sin t}{t}\, dt. \tag{12.189}$$

This is a Dirichlet integral. Its solution is $\pi/2$, such that

$$G(\mathbf{r}) = \frac{-1}{4\pi r}. \tag{12.190}$$

This is the formal solution of equation (12.173) without boundary conditions

$$T(\mathbf{x}) = \int \frac{-1}{4\pi} \frac{f(\mathbf{x}')}{|\mathbf{x} - \mathbf{x}'|}\, dv'. \tag{12.191}$$

Depending on the right-hand side $f(\mathbf{x})$, the integral to be solved may or may not have a simple closed-form expression.

12.8.2 The fixed-point iteration in real space as integral equation

Unfortunately, Green's function method cannot be applied directly to the boundary value problem of homogenization since it is a system of differential equations with nonconstant coefficients and a zero right-hand side $f(x)$. However, the eigenstrain problem has the suitable structure (see equation (8.7)), namely constant coefficients and a nonzero right side. In the case of the inhomogeneous heat conduction problem, we have

$$\nabla \cdot \mathbf{q} = 0, \quad \text{with } \mathbf{q}(\mathbf{x}) = \mathbf{L}^0 \mathbf{g}(\mathbf{x}) + \Delta\mathbf{L}(\mathbf{x})\mathbf{g}(\mathbf{x}) \quad \text{and} \quad \Delta\mathbf{L}(\mathbf{x}) = \mathbf{L}(\mathbf{x}) - \mathbf{L}^0. \tag{12.192}$$

Solving for $\mathbf{L}^0\mathbf{g}(\mathbf{x})$ and replacing $\mathbf{g}(\mathbf{x}) = \nabla T(\mathbf{x})$, this equation can be brought into a form that suits Green's approach,

$$\mathbf{L}^0 \nabla T(\mathbf{x}) = \mathbf{q}(\mathbf{x}) - \Delta\mathbf{L}(\mathbf{x})(\nabla T(\mathbf{x})). \tag{12.193}$$

We next take the divergence $\cdot\nabla$, which is

$$\mathbf{L}^0 : (\nabla T(\mathbf{x}) \otimes \nabla) = \mathbf{q}(\mathbf{x}) \cdot \nabla - \underbrace{[\Delta\mathbf{L}(\mathbf{x})(\nabla T(\mathbf{x}))]}_{\boldsymbol{\tau}(\mathbf{x})} \cdot \nabla. \tag{12.194}$$

The solution is $\mathbf{q}(\mathbf{x}) \cdot \nabla = 0$, such that

$$\mathbf{L}^0 : (\nabla T(\mathbf{x}) \otimes \nabla) = -[\Delta\mathbf{L}(\mathbf{x})(\nabla T(\mathbf{x}))] \cdot \nabla. \tag{12.195}$$

We now have the partial differential equation with constant coefficients on the left-hand side. The right-hand side is unknown since it contains the unknown function, but we can write down the solution $T(\mathbf{x})$ with Green's function anyway to obtain an integral equation for $T(\mathbf{x})$. This is a Fredholm integral of the second kind, and similar to the Lippmann–Schwinger-equation. For $\mathbf{L}^0 = l_0\mathbf{I}$ we obtain

$$\nabla^2 T(\mathbf{x}) = -l_0^{-1}[\Delta\mathbf{L}(\mathbf{x})(\nabla T(\mathbf{x}))] \cdot \nabla, \tag{12.196}$$

to which we can apply the Green solution of the Laplace operators,

$$T(\mathbf{x}) = \frac{1}{4\pi l_0} \int \frac{[\Delta\mathbf{L}(\mathbf{x}')(\nabla_{\mathbf{x}'} T(\mathbf{x}'))] \cdot \nabla_{\mathbf{x}'}}{|\mathbf{x} - \mathbf{x}'|} \,\mathrm{d}v'. \tag{12.197}$$

Since we are not interested in the temperature field but the effective relation between the average temperature gradient and the average heat flux, we take the gradient with $\nabla_{\mathbf{x}}$. This derivative can be pulled into the integral and is then applied to the integrand,

$$\mathbf{g}(\mathbf{x}) = \frac{1}{4\pi l_0} \int \frac{[\Delta\mathbf{L}(\mathbf{x}')\mathbf{g}(\mathbf{x}')] \cdot \nabla_{\mathbf{x}'}}{|\mathbf{x} - \mathbf{x}'|^3} (\mathbf{x} - \mathbf{x}') \,\mathrm{d}v'. \tag{12.198}$$

Repeated insertion of equation (12.198), i. e., using the integral as iterator, generates the correlation functions that we have seen in Section 3.3. With each application of the Green integral, a new location vector $\mathbf{x}'', \mathbf{x}''' \ldots$ appears. This results for discrete phases in integrals over products of indicator functions. In Milton (2002), Chapter 15, a more detailed presentation can be found.

Such iterations are interesting mainly due to their bounding properties. For certain choices of $\mathbf{L}_0$, one can show that the iteration converges from above or below.

A numerical evaluation of these integrals is somewhat cumbersome and cannot be recommended for several reasons. The integrals are singular (see the denominator in the integrand in equation (12.198)), and hence requires a renormalization (see Torquato, 1997) or a determination of the Cauchy principal value. Moreover, higher-order correlation functions are mostly not available. Additionally, the back transformation of the Green function to real space is not for all $\mathbf{L}^0$ possible in a closed form. This is problematic in elasticity theory when anisotropic stiffnesses are used; see, e. g., Section 5 in Mura (1987) for an overview.

13 Orientation averages

Most materials are anisotropic on the microscale. Typical examples are polycrystals, fiber-reinforced plastics, natural fibers as in wood and polymers with a semicrystalline structure. Only a few materials are isotropic on a molecular scale, like irregularly entangled molecule chains in rubber and glass-like substances.

Nevertheless, engineers employ isotropic material parameters. For example, for steel, the effective Young modulus is approximately 210 GPa and the effective Poisson ratio is near 0.3, although Young's modulus for single iron crystals can vary considerably depending on the tensile direction, namely between 280 GPa along the 111-direction and only 130 GPa along the 100-direction of the cubic crystal lattice. For a known orientation distribution and anisotropic single crystal properties, we can calculate with the aid of orientation averaging effective polycrystal properties. But in order to do so, we need to prepare some mathematical tools.

The following section is an introduction to rotations and orientations in 3D space. Good books can be found on this topic, of which I recommend Brannon (2018) and Popko (2012), which are somewhat approachable, and Morawiec (2004), which is more formal. Moreover, there are recordings of Professor V. Balakrishnan (the rotation group and all that, parts 1 to 3) available at https://www.youtube.com/watch?v=wIn_dlmD8sk, which are worth watching.

13.1 Orientations and rotations

SO(3) stands for the special orthogonal group over the 3D space. These are all linear transformations, i. e., 3×3-matrices, which preserve the scalar product between two vectors. Geometrically, only reflections and rotations have this property.

The group properties are intuitively clear: (1) The composition of two rotations remains a rotation (SO(3) is closed under composition), (2) to every rotation, there is a reversing rotation (each element has an inverse), (3) the identity **I** is contained in SO(3) as a rotation by zero degrees, (4) the matrix multiplication is associative. Elements of SO(3) are commonly abbreviated by **Q** or **R** (for rotation). With

$$\mathbf{a}' = \mathbf{Q} \cdot \mathbf{a} \tag{13.1}$$

$$\mathbf{b}' = \mathbf{Q} \cdot \mathbf{b} \tag{13.2}$$

it follows from the conservation of the scalar product

$$\mathbf{a}' \cdot \mathbf{b}' = \mathbf{a} \cdot \mathbf{b} = \mathbf{a} \cdot \mathbf{Q}^T \cdot \mathbf{Q} \cdot \mathbf{b}, \tag{13.3}$$

which holds only for arbitrary **a** and **b** if

$$\mathbf{Q}^T\mathbf{Q} = \mathbf{I}, \quad \mathbf{Q}^T = \mathbf{Q}^{-1}. \tag{13.4}$$

https://doi.org/10.1515/9783110793529-013

Therefore, we have $\det(\mathbf{Q}^T) = \det(\mathbf{Q}^{-1})$. Because of the determinant multiplication theorem, $\det(\mathbf{Q}^{-1}) = \det(\mathbf{Q})^{-1}$ and $\det(\mathbf{Q}) = \det(\mathbf{Q}^T)$, the determinant of must be $\det(\mathbf{Q}) = \pm 1$. The negative variant involves a central inversion $-\mathbf{I}$ or a mirror operation like $-\mathbf{e}_1 \otimes \mathbf{e}_1 + \mathbf{e}_2 \otimes \mathbf{e}_2 + \mathbf{e}_3 \otimes \mathbf{e}_3$, both having determinant -1. Here, we can neglect reflections, i. e., we restrict attention to rotations with $\det(\mathbf{Q}) = 1$, which is justified in Section 13.2.

From the tensorial equation (13.4), 6 independent scalar equations follow that any $\mathbf{Q}$ needs to satisfy. Hence, any $\mathbf{Q}$ can have only three independent components. The parameterization of SO(3) requires, therefore, three coordinates. We will see that a simple Cartesian parameterization of SO(3) is impossible.

An important property is the continuity of SO(3). Only then a parameterization with coordinates is possible. One speaks of a Lie group. This means that any group element is connected to any other group element by infinitesimal small, differential elements that are close to $\mathbf{I}$. All elements of a Lie-group can be obtained from a nonunique, minimal set of initial elements called the generators, which we will do now for SO(3).

Let us consider a rotation around the $\mathbf{e}_3$-axis:

$$\mathbf{Q}_{\omega\mathbf{e}_3} = \begin{bmatrix} \cos\omega & -\sin\omega & 0 \\ \sin\omega & \cos\omega & 0 \\ 0 & 0 & 1 \end{bmatrix} \mathbf{e}_i \otimes \mathbf{e}_j. \tag{13.5}$$

Note the right-hand rules, depicted in Figure 13.1: the first right-hand rule determines the orientation of the base vectors w. r. t. each other, and the second right-hand rule defines the positive sense of rotation.

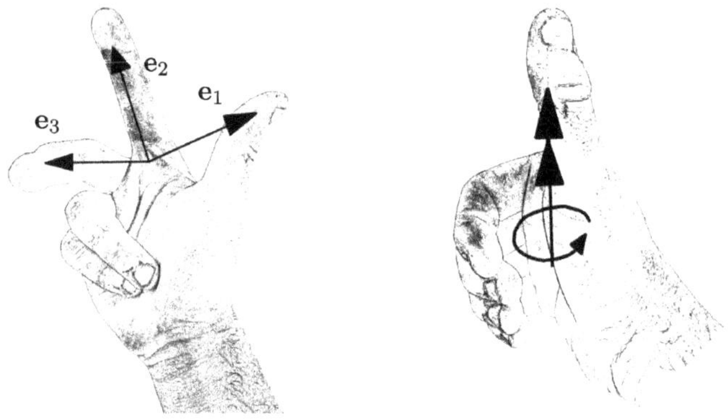

Figure 13.1: Both right-hand rules for defining a right-hand basis and the positive sense of rotation.

It is intuitively clear that we can divide the rotation into n smaller partial rotations around the angle ω/n. Linearizing at $\omega = 0$ with $\sin\omega/n \approx \omega/n$ and $\cos\omega/n \approx 1$ for small ω/n, we obtain

$$\mathbf{Q}_{\Delta\omega\mathbf{e}_3} \approx \mathbf{I} + \frac{\omega}{n} \underbrace{\begin{bmatrix} 0 & -1 & 0 \\ 1 & 0 & 0 \\ 0 & 0 & 0 \end{bmatrix} \mathbf{e}_i \otimes \mathbf{e}_j}_{\mathbf{W}_3}. \tag{13.6}$$

The tensor $\mathbf{W}_3$ is antisymmetric $\mathbf{W}_3^T = -\mathbf{W}_3$. It can be written with the aid of the permutation tensor and $\mathbf{e}_3$

$$\mathbf{W}_3 = -\overset{\langle 3 \rangle}{\boldsymbol{\varepsilon}} \cdot \mathbf{e}_3. \tag{13.7}$$

It acts on a vector $\mathbf{x}$ like the vector product with $\mathbf{e}_3$,

$$\mathbf{W}_3 \cdot \mathbf{x} = \mathbf{e}_3 \times \mathbf{x}. \tag{13.8}$$

Conversely, we have

$$\mathbf{e}_3 = -\frac{1}{2} \overset{\langle 3 \rangle}{\boldsymbol{\varepsilon}} : \mathbf{W}_3. \tag{13.9}$$

We are now ready to decompose the rotation $\mathbf{Q}_{\omega\mathbf{e}_3}$ into the product of partial rotations by the angle ω/n and then take the limit $n \to \infty$:

$$\begin{aligned}
\mathbf{Q}_{\omega\mathbf{e}_3} &= \prod_{i=1}^{n} \mathbf{Q}_{\omega/n\mathbf{e}_3} && (13.10)\\
&= (\mathbf{I} + \omega/n\, \mathbf{W}_3)^n && (13.11)\\
&= \binom{n}{0}\mathbf{I} + \binom{n}{1}\left(\frac{\omega}{n}\mathbf{W}_3\right)^1 + \dots + \binom{n}{n}\left(\frac{\omega}{n}\mathbf{W}_3\right)^n && (13.12)\\
&= \sum_{k=0}^{n} \binom{n}{k}\left(\frac{\omega}{n}\mathbf{W}_3\right)^k. && (13.13)
\end{aligned}$$

The binomial coefficients can be written as $\binom{n}{k} = n!/k!/(n-k)!$. We obtain for the n summands the expressions

$$\binom{n}{k}\left(\frac{\omega}{n}\mathbf{W}_3\right)^k = \frac{(n-k+1)(n-k+2)\dots(n)}{k!\,n^k}(\omega\mathbf{W}_3)^k \quad \leftarrow \quad \text{expand the numerator} \tag{13.14}$$

$$= \frac{n^k + \alpha n^{k-1} + \beta n^{k-2} + \dots + \zeta n^0}{k!\,n^k}(\omega\mathbf{W}_3)^k. \tag{13.15}$$

We can see now that in the case of $n \to \infty$, from the product in the denominator, only n^k remains, and all other products are of lower power. When the limit $n \to \infty$ is taken, only the expression $n^k/n^k/k!$ remains nonzero,

$$\lim_{n\to\infty} \frac{n^k + \alpha n^{k-1} + \beta n^{k-2} + \cdots + \zeta n^0}{k!\, n^k} (\omega \mathbf{W}_3)^k = \frac{1}{k!} (\omega \mathbf{W}_3)^k. \tag{13.16}$$

The entire sum can now be identified as the exponential function,

$$\mathbf{Q}_{\omega \mathbf{e}_3} = \lim_{n\to\infty} \sum_{k=0}^{n} \binom{n}{k} \left(\frac{\omega}{n} \mathbf{W}_3 \right)^k \tag{13.17}$$

$$= \sum_{k=0}^{n} \frac{1}{k!} (\omega \mathbf{W}_3)^k \tag{13.18}$$

$$= \exp(\omega \mathbf{W}_3). \tag{13.19}$$

The application of the exponential function is most conveniently carried out in the spectral representation. $\omega \mathbf{W}_3$ has the following eigenvalues and right and left eigenvectors:

$$\lambda_1 = \mathrm{i}\omega \qquad \mathbf{r}_1 = \mathrm{i}\mathbf{e}_1 + \mathbf{e}_2 \qquad \mathbf{l}_1 = -\mathrm{i}\mathbf{e}_1 + \mathbf{e}_2 \tag{13.20}$$

$$\lambda_2 = -\mathrm{i}\omega \qquad \mathbf{r}_2 = -\mathrm{i}\mathbf{e}_1 + \mathbf{e}_2 \qquad \mathbf{l}_2 = \mathrm{i}\mathbf{e}_1 + \mathbf{e}_2 \tag{13.21}$$

$$\lambda_3 = 0 \qquad \mathbf{r}_3 = \mathbf{e}_3 \qquad \mathbf{l}_3 = \mathbf{e}_3. \tag{13.22}$$

In the spectral decomposition, the exponential function is

$$\mathbf{Q}_{\omega \mathbf{e}_3} = \sum_{i=1}^{3} \frac{\exp(\lambda_i)}{\mathbf{l}_i \cdot \mathbf{r}_i} \mathbf{r}_i \otimes \mathbf{l}_i. \tag{13.23}$$

The division by the scalar product servers for normalizing the eigenvectors. It remains

$$\exp(\omega \mathbf{W}_3) = \frac{\exp(\mathrm{i}\omega)}{2} \begin{array}{c|ccc} & -\mathrm{i} & 1 & 0 \\ \hline \mathrm{i} & & & \\ 1 & & & \\ 0 & & & \end{array} + \frac{\exp(-\mathrm{i}\omega)}{2} \begin{array}{c|ccc} & \mathrm{i} & 1 & 0 \\ \hline -\mathrm{i} & & & \\ 1 & & & \\ 0 & & & \end{array} + \exp(0) \begin{array}{c|ccc} & 0 & 0 & 1 \\ \hline 0 & & & \\ 0 & & & \\ 1 & & & \end{array} \tag{13.24}$$

$$= \frac{\exp(\mathrm{i}\omega)}{2} \begin{bmatrix} 1 & \mathrm{i} & 0 \\ -\mathrm{i} & 1 & 0 \\ 0 & 0 & 0 \end{bmatrix} + \frac{\exp(-\mathrm{i}\omega)}{2} \begin{bmatrix} 1 & -\mathrm{i} & 0 \\ \mathrm{i} & 1 & 0 \\ 0 & 0 & 0 \end{bmatrix} + \begin{bmatrix} 0 & 0 & 0 \\ 0 & 0 & 0 \\ 0 & 0 & 1 \end{bmatrix}. \tag{13.25}$$

With the rule $\exp(\pm\mathrm{i}\omega) = \cos\omega \pm \mathrm{i}\sin\omega$, we can summarize further

$$\mathbf{Q}_{\omega \mathbf{e}_3} = \frac{\cos\omega + \mathrm{i}\sin\omega}{2} \begin{bmatrix} 1 & \mathrm{i} & 0 \\ -\mathrm{i} & 1 & 0 \\ 0 & 0 & 0 \end{bmatrix} + \frac{\cos\omega - \mathrm{i}\sin\omega}{2} \begin{bmatrix} 1 & -\mathrm{i} & 0 \\ \mathrm{i} & 1 & 0 \\ 0 & 0 & 0 \end{bmatrix} + \begin{bmatrix} 0 & 0 & 0 \\ 0 & 0 & 0 \\ 0 & 0 & 1 \end{bmatrix} \tag{13.26}$$

$$= \begin{bmatrix} \cos\omega & -\sin\omega & 0 \\ \sin\omega & \cos\omega & 0 \\ 0 & 0 & 1 \end{bmatrix}. \tag{13.27}$$

One can see that $\mathbf{Q}_{\omega\mathbf{e}_3}$ can be written symbolically as

$$\mathbf{Q}_{\omega\mathbf{e}_3} = \cos\omega(\mathbf{I} - \mathbf{e}_3 \otimes \mathbf{e}_3) - \sin\omega \overset{\langle 3 \rangle}{\boldsymbol{\varepsilon}} \cdot \mathbf{e}_3 + \mathbf{e}_3 \otimes \mathbf{e}_3. \tag{13.28}$$

The generalization for arbitrary axes of rotation $\mathbf{r}$ is obtained by replacing $\mathbf{e}_3$ by $\mathbf{r}$ in the latter equation, where $\mathbf{r}$ needs to be normalized. A geometric interpretation of the above equation (13.28) is depicted in Figure 13.2. We can now discuss methods to parameterize SO(3) by three parameters.

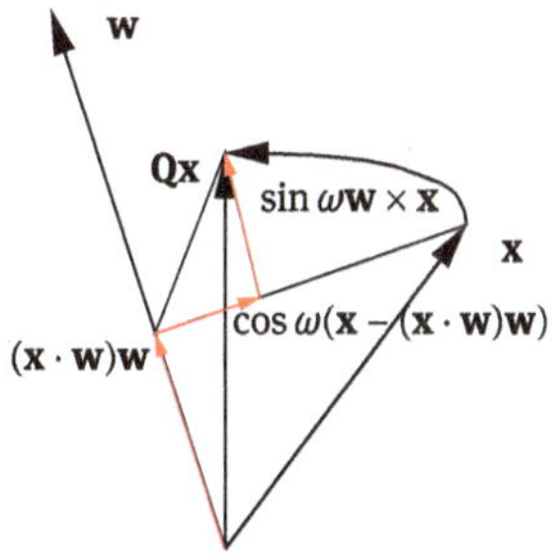

Figure 13.2: Geometric representation of the formula for finite rotations (equation (13.28)).

13.1.1 Axis-angle parameterization

A first obvious approach is to parameterize the direction of the rotation axis by two angles, for example, the longitude ϕ and the latitude θ,

$$\mathbf{w} = \cos\phi\cos\theta\mathbf{e}_1 + \sin\phi\cos\theta\mathbf{e}_2 + \sin\theta\mathbf{e}_3. \tag{13.29}$$

This is used to construct an antisymmetric tensor $-\overset{\langle 3 \rangle}{\boldsymbol{\varepsilon}} \cdot \mathbf{w} = \mathbf{W}$ which is multiplied by the angle of rotation ω. The exponential function applied to $\mathbf{W}$ gives the sought orthogonal rotation tensor. The explicit representation yields the Rodrigues[1] formula for finite rotations

$$\mathbf{Q}_{\omega\mathbf{w}} = \exp(-\omega \overset{\langle 3 \rangle}{\boldsymbol{\varepsilon}} \cdot \mathbf{w}) \tag{13.30}$$

$$= \cos\omega(\mathbf{I} - \mathbf{w} \otimes \mathbf{w}) - \sin\omega \overset{\langle 3 \rangle}{\boldsymbol{\varepsilon}} \cdot \mathbf{w} + \mathbf{w} \otimes \mathbf{w}. \tag{13.31}$$

We see that due to the normality $|\mathbf{w}| = 1$ and the traceless property $W_{ii} = 0$,

1 Olinde Rodrigues, 1795–1851 (https://en.wikipedia.org/wiki/Olinde_Rodrigues).

$$\mathrm{sp}(\mathbf{Q}_{\omega\mathbf{w}}) = 2\cos\omega + 1 \tag{13.32}$$

holds, such that

$$\omega = \arccos\left(\frac{\mathrm{sp}(\mathbf{Q}_{\omega\mathbf{w}}) - 1}{2}\right). \tag{13.33}$$

Moreover, we see that rotations by 180° result in the simple expression

$$\mathbf{Q}_{\pi\mathbf{w}} = -\mathbf{I} + 2\mathbf{w} \otimes \mathbf{w}. \tag{13.34}$$

Summarizing ω with $\mathbf{w}$ gives an illustrative geometric representation of SO(3). Every point in a sphere with a radius π represents a rotation, where the directions from the origin are the rotation axes and the distances from the origin are the rotation angles.

The radius is restricted to $0 \leq \omega \leq \pi$ due to the periodicity of rotations. Rotations by 180° form the boundary of the sphere and are contained twice at antipodal points.

This makes SO(3) an interesting topological space. A closed path that crosses the surface of 180° rotations *once* cannot be contracted to a point similar to a round trip on a torus. A closed path that crosses the surface of 180° rotations *twice* can be contracted to a point, as sketched in Figure 13.3. This observation can be extended to all closed paths with an odd and an even number of crossings of the 180° surface. SO(3) is three-dimensional, but we cannot construct a body with such a topology. SO(3) is said to be doubly connected since we can construct two kinds of closed paths that can be contracted to points, namely paths without crossings and with an even number of crossings of the 180° surface. Therefore, a Cartesian parameterization of SO(3) is impossible.

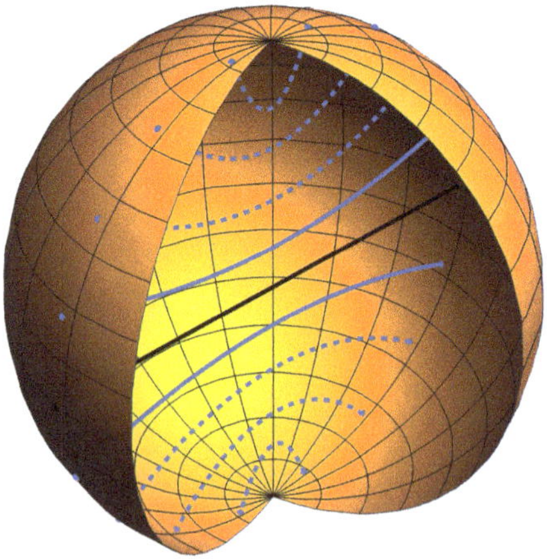

Figure 13.3: The black line represents a simple cyclic path through SO(3) that cannot be contracted to a point. The blue lines represent a double cyclic path through SO(3) that can be contracted to a point, although it can be arbitrarily close to the simple cyclic path.

13.1.2 Euler angle parameterization

Every rotation can be obtained as a composition of three elementary rotations around fixed axes. One might, for example, construct and use $\mathbf{W}_1$ and $\mathbf{W}_2$ analogous to $\mathbf{W}_3 = -\overset{\langle 3\rangle}{\boldsymbol{\varepsilon}} \cdot \mathbf{e}_3$. The order of composition is not arbitrary. The rules for calculating with the exponential function are not entirely transferable from real calculus,

$$\begin{aligned} \mathbf{Q} &= \mathbf{Q}_{\phi_1 \mathbf{e}_1} \mathbf{Q}_{\phi_2 \mathbf{e}_2} \mathbf{Q}_{\phi_3 \mathbf{e}_3} && (13.35)\\ &= \exp(\phi_1 \mathbf{W}_1) \exp(\phi_2 \mathbf{W}_2) \exp(\phi_3 \mathbf{W}_3) && (13.36)\\ &\neq \exp(\phi_1 \mathbf{W}_1 + \phi_2 \mathbf{W}_2 + \phi_3 \mathbf{W}_3). && (13.37) \end{aligned}$$

Conventions

Including the order of composition and the possibility to chose spatial fixed axes or rotating axes, a variety of possible combinations emerges. The most common choice is the $z-x-z$-convention with spatially fixed axes. The first rotation is by ϕ_1 around $\mathbf{e}_3$. The second rotation is by Φ around $\mathbf{e}_1$. The third rotation is by ϕ_2 again around $\mathbf{e}_3$,

$$\mathbf{Q} = \mathbf{Q}_{\phi_2 \mathbf{e}_3} \mathbf{Q}_{\Phi \mathbf{e}_1} \mathbf{Q}_{\phi_1 \mathbf{e}_3}. \tag{13.38}$$

The advantage of this choice is that one can construct inverse easily by changing the signs of the angles and interchanging ϕ_1 and ϕ_2.

Advantages

All Euler[2] angle parameterizations have in common that $\mathbf{Q}$ can be easily denoted with respect to the orthonormal basis $\mathbf{e}_i \otimes \mathbf{e}_j$. In the $z-x-z$-convention that is adopted here, we have

$$\begin{aligned} \mathbf{Q} &= \mathbf{Q}_{\phi_2 \mathbf{e}_3} \mathbf{Q}_{\Phi \mathbf{e}_1} \mathbf{Q}_{\phi_1 \mathbf{e}_3} && (13.39)\\ &= \begin{bmatrix} c_2 & -s_2 & 0 \\ s_2 & c_2 & 0 \\ 0 & 0 & 1 \end{bmatrix} \cdot \begin{bmatrix} 1 & 0 & 0 \\ 0 & c & -s \\ 0 & s & c \end{bmatrix} \cdot \begin{bmatrix} c_1 & -s_1 & 0 \\ s_1 & c_1 & 0 \\ 0 & 0 & 1 \end{bmatrix} \\ &= \begin{bmatrix} c_1c_2 - cs_1s_2 & -c_2s_1 - cc_1s_2 & ss_2 \\ cc_2s_1 + c_1s_2 & cc_1c_2 - s_1s_2 & -c_2s \\ ss_1 & c_1s & c \end{bmatrix} && (13.40) \end{aligned}$$

with

$$\begin{aligned} s_1 &= \sin\phi_1 & s_2 &= \sin\phi_2 & s &= \sin\Phi && (13.41)\\ c_1 &= \cos\phi_1 & c_2 &= \cos\phi_2 & c &= \cos\Phi. && (13.42) \end{aligned}$$

2 Leonhard Euler, 1707–1783 (https://en.wikipedia.org/wiki/Leonhard_Euler).

Integrations over SO(3) in Euler angles are then reduced to integrations over sine and cosine functions, which can, in general, be solved easily.

Drawbacks

Euler parameterizations are not unique for rotations around $\mathbf{e}_3$, which can be split arbitrarily between ϕ_1 and ϕ_2. Regarding the build-in periodicity of SO(3), the parameterization can be made unique by restricting the range of definition of the angles, specifically

$$0 \le \phi_1 < 2\pi \tag{13.43}$$

$$0 \le \Phi \le \pi \tag{13.44}$$

$$0 \le \phi_2 < 2\pi. \tag{13.45}$$

The restriction of Φ to the interval $0 \ldots \pi$ is necessary, otherwise two parameter sets for ϕ_1, Φ, ϕ_2 represent the same rotation, namely

$$\phi'_{1,2} = \phi_{1,2} + \pi \tag{13.46}$$

$$\Phi' = 2\pi - \Phi. \tag{13.47}$$

The transition to the $'$-angles just changes the signs of $s_{1,2}$, $c_{1,2}$ and s. Reviewing the matrix in equation (13.40), one can see that these changes of sign cancel out, i. e., they give the same rotation matrix.

Further, sequential elementary Euler rotations are efficient representations with respect to the basis $\mathbf{e}_i \otimes \mathbf{e}_j$, but they can be an inefficient detour in SO(3); see the *robot dance* Section (15.8.1) in Brannon (2018).

13.1.3 Rodrigues parameterization

Let $\mathbf{Q}$ be a rotation matrix that maps $\mathbf{x}$ to $\mathbf{x}' = \mathbf{Q}\mathbf{x}$. Let $\mathbf{y}$ and $\mathbf{z}$ be auxiliary quantities depicted in Figure 13.4,

$$\mathbf{y} = \mathbf{x} - \mathbf{x}' = (\mathbf{I} - \mathbf{Q})\mathbf{x} \tag{13.48}$$

$$\mathbf{z} = \mathbf{x} + \mathbf{x}' = (\mathbf{I} + \mathbf{Q})\mathbf{x}. \tag{13.49}$$

We can express $\mathbf{y}$ in terms of $\mathbf{z}$,

$$\mathbf{y} = (\mathbf{I} - \mathbf{Q})\mathbf{x} \tag{13.50}$$

$$= \underbrace{(\mathbf{I} - \mathbf{Q})(\mathbf{I} + \mathbf{Q})^{-1}}_{\mathbf{W}_\mathrm{R}}\mathbf{z}, \tag{13.51}$$

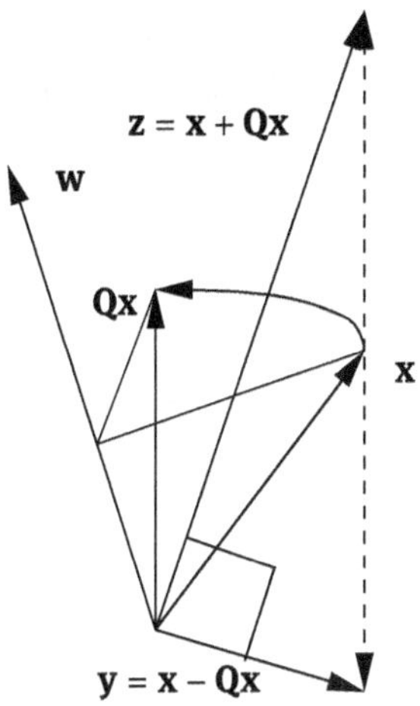

Figure 13.4: Sketch of the auxiliary quantities $\mathbf{z} = \mathbf{x} + \mathbf{Q}\mathbf{x}$ and $\mathbf{y} = \mathbf{x} - \mathbf{Q}\mathbf{x}$ for deriving the Rodrigues parameterization.

where we assume the invertibility of $\mathbf{I} + \mathbf{Q}$. This holds for all rotations around angles different from zero and multiples of 180°. One sees that $\mathbf{z}$ and $\mathbf{y}$ are perpendicular to each other; hence,

$$\mathbf{z} \cdot \mathbf{y} = \mathbf{z} \cdot \mathbf{W}_\mathrm{R} \cdot \mathbf{z} = 0. \tag{13.52}$$

This needs to hold for arbitrary $\mathbf{x}$, and with the assumed invertibility, for all $\mathbf{z}$. Therefore, $\mathbf{W}_\mathrm{R}$ must be antisymmetric. The antisymmetric tensor acts as if a cross-product is taken therefore $\mathbf{W}_\mathrm{R}\mathbf{z}$ is perpendicular to $\mathbf{z}$. The symmetric part has only real eigenvalues and real, orthogonal eigendirections, such that $\mathbf{z} \cdot \mathbf{W}_\mathrm{R} \cdot \mathbf{z}$ is only zero for all vectors $\mathbf{z}$ if the symmetric part is zero. We can solve

$$\mathbf{W}_\mathrm{R} = (\mathbf{I} - \mathbf{Q})(\mathbf{I} + \mathbf{Q})^{-1} \tag{13.53}$$

for $\mathbf{Q}$ by multiplying with $\mathbf{I} + \mathbf{Q}$, bring all terms with $\mathbf{Q}$ to one side, factor out $\mathbf{Q}$ and then isolate $\mathbf{Q}$ on one side,

$$\mathbf{Q} = (\mathbf{I} + \mathbf{W}_\mathrm{R})^{-1}(\mathbf{I} - \mathbf{W}_\mathrm{R}). \tag{13.54}$$

The antisymmetric tensor $\mathbf{W}_\mathrm{R}$ can be represented by three independent parameters,

$$\mathbf{W}_\mathrm{R} = -\overset{\langle 3 \rangle}{\boldsymbol{\varepsilon}} \cdot \mathbf{w}_\mathrm{R}. \tag{13.55}$$

This is another apparently nonperiodic parameterization of $\mathbf{Q}$. Interestingly, $\mathbf{I} + \mathbf{W}_\mathrm{R}$ is always invertible, which can only imply that not the entire SO(3) space is covered since 180° rotations need to appear twice. Figuratively, the Rodrigues parameterization corresponds to inflating the SO(3)-sphere of the axis-angle parameterization in

Figure 13.3 by expanding the 180° boundary to $+\infty$. This removes the periodicity at the cost of not covering 180° rotations, which are contained only as directional limits for $\| \pm \mathbf{w}_\mathrm{R}\| \to \infty$. Then the eigenvalues of $\mathbf{I}$ vanish compared to the eigenvalues of $\mathbf{W}_\mathrm{R}$, such that $\mathbf{I} + \mathbf{W}_\mathrm{R}$ is not invertible.

13.2 Rotations of tensors of arbitrary order

The rotation of a vector $\mathbf{v}$ is

$$\mathbf{v}' = \mathbf{Q} \cdot \mathbf{v} = Q_{ij}\mathbf{e}_i \otimes \mathbf{e}_j \cdot v_k \mathbf{e}_k = \underbrace{Q_{ij}v_j}_{v_i'}\,\mathbf{e}_i = v_j\,\underbrace{Q_{ij}\mathbf{e}_i}_{\mathbf{e}_j'}\,. \tag{13.56}$$

Mostly the products are summarized as $v_i' = Q_{ij}v_j$, which is the new component w. r. t. the basis $\mathbf{e}_i$. Alternatively, one can associate $Q_{ij}\mathbf{e}_i = \mathbf{e}_j'$ and consider this the rotated basis to the unaltered component v_j. The latter can be generalized easily to tensors of arbitrary order s. To do so, we define the Rayleigh product

$$\overset{\langle s\rangle}{\mathbb{A}}{}' = \mathbf{Q} * \overset{\langle s\rangle}{\mathbb{A}} = A_{ij\ldots MN}(\mathbf{Q} \cdot \mathbf{e}_i) \otimes \cdots \otimes (\mathbf{Q} \cdot \mathbf{e}_n), \tag{13.57}$$

with the components

$$A_{ij\ldots mn} = \overset{\langle s\rangle}{\mathbb{A}} \cdot \cdots \cdot \mathbf{e}_i \otimes \cdots \otimes \mathbf{e}_n. \tag{13.58}$$

It has the following properties:

- It is homogeneous of degree s in the first factor, $(\alpha\mathbf{Q}) * \overset{\langle s\rangle}{\mathbb{A}} = \alpha^s(\mathbf{Q} * \overset{\langle s\rangle}{\mathbb{A}})$. Orthogonal tensors that involve reflections or central inversions differ from rotations by the factor -1. Because we will only rotate tensors of even order s we have $(-1)^s = 1$, i. e., the reflections of even-ordered tensors cannot be distinguished from rotations and need not be considered explicitly.
- It is associative in the first factor w. r. t. multiplication,

$$(\mathbf{Q}_1\mathbf{Q}_2) * \overset{\langle s\rangle}{\mathbb{A}} = \mathbf{Q}_1 * (\mathbf{Q}_2 * \overset{\langle s\rangle}{\mathbb{A}}). \tag{13.59}$$

- It is linear in the second factor. This allows to write the Rayleigh product as a linear mapping with a tensor of order $2s$,

$$\overset{\langle s\rangle}{\mathbb{A}}{}' = \overset{\langle 2s\rangle}{\mathbb{Q}}(\mathbf{Q}) \underbrace{\cdot\cdot\cdot\cdot}_{s\text{ scalar dots}} \overset{\langle s\rangle}{\mathbb{A}}. \tag{13.60}$$

  In this representation, the linearity is more apparent and can be easily joined with other linear operations.

This is a nice example of how the abstract tensor notation allows to easily generalize rotations from vectors, a task next to impossible if the basis is omitted, for example, in matrix calculus.

13.3 Hooke's anisotropic law

Following the Neumann[3]–Curie[4] principle, spatial symmetries of the structure of a material manifest in the physical properties of the material (Neumann, 1885; Curie, 1894). For example, if an atomic lattice is invariant under the rotation $\mathbf{Q}$, likewise is the stiffness $\mathbb{C}$ invariant under this rotation,

$$\mathbb{C} = \mathbf{Q} * \mathbb{C}. \tag{13.61}$$

Spatial symmetries are mathematically represented by symmetry groups $\mathcal{G} = \{\mathbf{Q}_1, \mathbf{Q}_2 \dots \mathbf{Q}_n\}$, which are subgroups of SO(3). One can find a general analysis of the subgroups of SO(3) in Weyl (1939), Auffray et al. (2019). The group axioms are associativity, closedness, the existence of an inverse element to all group members and the identity $\mathbf{I}$ as a group member,

$$(\mathbf{Q}_1\mathbf{Q}_2)\mathbf{Q}_3 = \mathbf{Q}_1(\mathbf{Q}_2\mathbf{Q}_3) = \mathbf{Q}_4, \quad \mathbf{Q}_{1,2,3,4} \in \mathcal{G} \tag{13.62}$$

$$\mathbf{Q}_1\mathbf{Q}_1^{-1} = \mathbf{I}, \quad \mathbf{Q}_1, \mathbf{Q}_1^{-1}, \mathbf{I} \in \mathcal{G}. \tag{13.63}$$

The group elements can be generated by a nonunique, minimal set of generators $\{\mathbf{Q}_{G1}, \mathbf{Q}_{G2}, \dots \mathbf{Q}_{Gm}\}$ with $m \le n$. The symmetry constraints are enforced by requiring

$$\mathbb{C} = \mathbf{Q}_i * \mathbb{C} \quad \forall \mathbf{Q}_i \in \mathcal{G}. \tag{13.64}$$

Due to the associativity of the Rayleigh product in the first factor and the representation of all $\mathbf{Q}_i$ by, for example, two generators $\mathbf{G}_{G1,G2}$,

$$\mathbb{C} = \mathbf{Q}_i * \mathbb{C} \tag{13.65}$$

$$= (\mathbf{Q}_{G1}\mathbf{Q}_{G2} \dots \mathbf{Q}_{G1}) * \mathbb{C} \tag{13.66}$$

$$= \mathbf{Q}_{G1} * (\mathbf{Q}_{G2} * \dots (\mathbf{Q}_{G1} * \mathbb{C})), \tag{13.67}$$

every symmetry transformation $\mathbf{Q}_i$ can be reduced to the action of the generators $\mathbf{Q}_{G1,G2}$. Therefore, equation (13.64) needs to be evaluated only for one set of generators. Moreover, equation (13.61) is a linear, homogeneous system of equations for $\mathbb{C}$, since the Rayleigh product is linear in the second factor,

3 Franz Ernst Neumann, 1798–1895 (https://en.wikipedia.org/wiki/Franz_Ernst_Neumann).

4 Pierre Curie, 1859–1906 (https://en.wikipedia.org/wiki/Pierre_Curie).

$$\mathbb{C} = \mathbf{Q}_i * \mathbb{C} \tag{13.68}$$

$$\mathbf{Q}_i * \mathbb{C} - \mathbb{C} = \mathbb{O} \tag{13.69}$$

$$\underbrace{(\overset{\langle 8\rangle}{\mathbb{Q}} - \overset{\langle 8\rangle}{\mathbb{I}})}_{\mathbb{L}} :: \mathbb{C} = \mathbb{O} \quad \text{with } \overset{\langle 8\rangle}{\mathbb{Q}} = Q_{im}Q_{jn}Q_{ko}Q_{lp}\mathbf{e}_i \otimes \mathbf{e}_j \otimes \mathbf{e}_k \otimes \mathbf{e}_l \otimes \mathbf{e}_m \otimes \mathbf{e}_n \otimes \mathbf{e}_o \otimes \mathbf{e}_p. \tag{13.70}$$

The solution space for $\mathbb{C}$ depends on $\mathbb{L}$. Due to the fact that $\mathbf{Q}$ appears with the power four and since powers n of trigonometric functions can be written as linear combinations of trigonometric functions with periods $2\pi/k$ with $k = 1 \dots n$,

$$\sin^n x = \frac{1}{2^n} \sum_{k=0}^{n} \binom{n}{k} \cos((n-2k)(x-\pi/2)), \tag{13.71}$$

$$\cos^n x = \frac{1}{2^n} \sum_{k=0}^{n} \binom{n}{k} \cos((n-2k)(x)). \tag{13.72}$$

Interesting $\mathbb{L}$ are obtained only for $2\pi/k$ with $k = 1 \dots 4$. For such angles, a rank reduction of $\mathbb{L}$ occurs, such that the number of linear constraints is reduced, and the number of independent components of $\mathbb{C}$ is bigger.

For higher-fold rotations with $k = 5 \dots \infty$, no further rank reduction occurs, such that all higher-fold rotations coincide with the continuous rotational symmetry (transversal isotropy). This has been described by Hermann (1934),[5] and is demonstrated in Listing 13.2.

This is a noteworthy observation, which implies that linear laws can only distinguish a limited number of symmetries, depending on the order of the constitutive tensor. For example, linear heat conduction involves a second-order constitutive tensor. In this case, only five symmetries can be distinguished, and cubic symmetry can not be distinguished from isotropy. Measurements, however, show that for temperatures below approximately 30 Kelvin, heat conduction becomes anisotropic in cubic crystals (McCurdy et al., 1970). Linear laws are then not applicable.

In linear elasticity, eight symmetry classes can be distinguished; see Table 13.1. The largest symmetry group symmetry group is SO(3), and the smallest one is the triclinic group $\{\mathbf{I}\}$. We consider the latter as a trivial example: No restrictions can be derived for the components of $\mathbb{C}$. Therefore, triclinic material behavior is described by all $21 = 1 + 2 + 3 + 4 + 5 + 6$ independent components of a symmetric 6×6-matrix in the Voigt–Mandel notation.

A decomposition of the stiffness into an orientation part (characterized by at most three orientation parameters) and a moduli part (the actual anisotropic stiffnesses) is recommended, as described in Kowalczyk-Gajewska and Ostrowska-Maciejewska

5 Carl Hermann, 1898–1961 (https://en.wikipedia.org/wiki/Carl_H._Hermann).

Table 13.1: The symmetry groups of elasticity. In the third column, the number of classically given free parameters is given, which is obtained when the orientations for specifying the generators are neglected. For example, the difference 15 − 13 = 2 for monoclinic symmetry is explained by the missing two angles that specify the rotation axis of the generator $\mathbf{Q}_{\mathbf{e}_1}^{\pi}$. For triclinic symmetry, no parameters are needed for specifying the generators. For transversal isotropy and n-gonal symmetry, two parameters are sufficient to give the direction of the rotation axis. The differential generators in the case of isotropy are arbitrary and require no parameters. Orthotropic, cubic, trigonal and tetragonal generators need to be specified by three parameters.

symmetry	generators (example)	eigenvalues + coupling parameters + ori.-parameters = indep. comps. in $\mathbb{C}$	$\lvert\mathcal{G}\rvert$
triclinic	$\mathbf{I}$	$6 + 12 + 3 = 21$ (21)	1
monoclinic	$\mathbf{Q}_{\mathbf{e}_1}^{\pi}$	$6 + 6 + 3 = 15$ (13)	2
orthotropic	$\mathbf{Q}_{\mathbf{e}_1}^{\pi}, \mathbf{Q}_{\mathbf{e}_2}^{\pi}$	$6 + 3 + 3 = 12$ (9)	4
trigonal	$\mathbf{Q}_{\mathbf{e}_2}^{\pi}, \mathbf{Q}_{\mathbf{e}_3}^{2\pi/3}$	$4 + 2 + 3 = 9$ (6)	6
tetragonal	$\mathbf{Q}_{\mathbf{e}_1}^{\pi}, \mathbf{Q}_{\mathbf{e}_3}^{2\pi/4}$	$4 + 2 + 3 = 9$ (6)	8
n-gonal	$\mathbf{Q}_{\mathbf{e}_1}^{\pi}, \mathbf{Q}_{\mathbf{e}_3}^{2\pi/n}, n > 4$	$4 + 1 + 2 = 7$ (5)	$2n$
trans.-iso.	$\mathbf{Q}_{\mathbf{e}_1}^{\pi}, \mathbf{Q}_{\mathbf{e}_3}^{\mathrm{d}\phi}$	$4 + 1 + 2 = 7$ (5)	∞
cubic	$\mathbf{Q}_{\mathbf{e}_1}^{3\pi/2}, \mathbf{Q}_{\frac{1}{\sqrt{3}}(\mathbf{e}_1+\mathbf{e}_2+\mathbf{e}_3)}^{2\pi/3}$	$3 + 0 + 3 = 6$ (3)	24
isotropic	2 diff. generators $\mathbf{Q}_{\mathbf{e}_1}^{\mathrm{d}\phi}, \mathbf{Q}_{\mathbf{e}_2}^{\mathrm{d}\theta}$ see Section 13.5	$2 + 0 + 0 = 2$ (2)	∞

(2009). The orientation information of a triclinic cell requires three angles. The moduli information is then contained in the spectral representation of $\mathbb{C}$; see Table 1 in Cowin and Mehrabadi (1992). For triclinic material, one finds six eigenvalues (Kelvin moduli) and 12 coupling parameters (also referred to as *elastic distributors*), which relate the eigenprojectors with respect to each other. An overview is given in Table 13.1. In Figure 13.5, the subset relations among the symmetry classes are depicted.

It is noteworthy that the numbers of independent components that are usually given in the literature include the orientation information inconsistently. For example, the 21 independent components include the orientation information ($21 = 18 + 3$), the 13 independent components for monoclinic symmetry contain only one orientation parameter ($12 + 1 = 13$), although three orientation parameters are needed, and all other classically given numbers of independent components contain no orientation parameter. The classically given numbers of independent components of $\mathbb{C}$ is given in Table 13.1 in column three in parenthesis (see, e. g., Ting, 1996).

In Listing 13.1, it is shown how this information is obtained with the aid of a computer algebra system. The example for cubic symmetry shows that the symmetry group contains 24 elements, for fixed generators, three free moduli remain in $\mathbb{C}$ and that one possible representation of $\mathbb{C}$ is with respect to the same cubic axes with the free components C_{2222}, C_{2323} and C_{2233}.

Listing 13.1: Mathematica notebook (https://gitlab.com/gluegerainer/listings-homogenization-methods/-/blob/main/Listing-27_en.nb) for generating symmetry groups from the generators and to reduce the stiffness $\mathbb{C}$ to independent components.

```
Remove["Global`*"]
(* The symmetry group consists initially only of its generators *)
id = IdentityMatrix[3];
List1 = {id};                                                          (* triclinic *)
List2 = {-id+{{2,0,0},{0,0,0},{0,0,0}}};                               (* monoclinic *)
List3 = {-id+{{2,0,0},{0,0,0},{0,0,0}},-id+{{0,0,0},{0,2,0},{0,0,0}}}; (* orthotropic *)
List4 = {RotationMatrix[2Pi/3,{0,0,1}],RotationMatrix[Pi,{1,0,0}]};    (* trigonal *)
List5 = {RotationMatrix[Pi/2,{0,0,1}],-id+{{2,0,0},{0,0,0},{0,0,0}}};  (* tetragonal *)
List6 = {RotationMatrix[Pi/3,{0,0,1}],-id+{{2,0,0},{0,0,0},{0,0,0}}};  (* hexagonal *)
List7 = {RotationMatrix[Pi/2,{1,0,0}],RotationMatrix[2Pi/3,{1,1,1}]};  (* cubic *)

(* Example for cubic symmetry *)
List = List7;
init = Length[List];
outit = 100000;
(* Generate elements until no new elements appear *)
While[
 init != outit,
 init = Length[List];
 For[i = 1, i <= init, i += 1,
  For[j = 1, j <= init, j += 1,
    AppendTo[List, FullSimplify[List[[i]].List[[j]]]]
    ];
  ];
 List = DeleteDuplicates[List];
 outit = Length[List];
 Print["Number of elements in G: ", outit]]

(* Create general fourth-order tensor *)
c=Table[ToExpression[StringJoin["C",ToString[i],ToString[j],ToString[k],ToString[l]]],
    {i,1,3},{j,1,3},{k,1,3},{l,1,3}];
(* Require rotational symmetries *)
Q=List[[1]];
eqs1=Table[c[[m,n,o,p]]==Sum[c[[i,j,k,l]]Q[[m,i]] Q[[n,j]] Q[[o,k]] Q[[p,l]],
    {i,1,3},{j,1,3},{k,1,3},{l,1,3}],{m,1,3},{n,1,3},{o,1,3},{p,1,3}];
Q=List[[2]];
eqs2=Table[c[[m,n,o,p]]==Sum[c[[i,j,k,l]] Q[[m,i]] Q[[n,j]] Q[[o,k]] Q[[p,l]],
    {i,1,3},{j,1,3},{k,1,3},{l,1,3}],{m,1,3},{n,1,3},{o,1,3},{p,1,3}];
(* Subsymmetries and principle symmetry *)
eqs4=Table[c[[m,n,o,p]]==c[[m,n,p,o]],{m,1,3},{n,1,3},{o,1,3},{p,1,3}];
eqs5=Table[c[[m,n,o,p]]==c[[o,p,m,n]],{m,1,3},{n,1,3},{o,1,3},{p,1,3}];
(* Summarize equations *)
eqs=Flatten[{eqs1,eqs2,eqs4,eqs5}];
(* 3^4 minus the number of independent equations (rank of the coefficient matrix) is the *)
(* number of independent components w.r.t. the symmetry group. *)
Print["Number of independent components: ", 81-MatrixRank[CoefficientArrays[eqs,Flatten[c
      ]][[2]]]]

(* Show stiffness without dependent components. *)
erg = Solve[eqs, Flatten[c]];
Set @@@ erg[[1]];
c
-----------------------------------------
Result:
Number of elements in G: 6
Number of elements in G: 18
```

```
Number of elements in G: 24
Number of elements in G: 24
Number of independent components: 3
{{{{C2222,0,0},{0,C2233,0},{0,0,C2233}},
{{0,C2332,0},{C2332,0,0},{0,0,0}},
{{0,0,C2332},{0,0,0},{C2332,0,0}}},
{{{0,C2332,0},{C2332,0,0},{0,0,0}},
{{C2233,0,0},{0,C2222,0},{0,0,C2233}},
{{0,0,0},{0,0,C2332},{0,C2332,0}}},
{{{0,0,C2332},{0,0,0},{C2332,0,0}},
{{0,0,0},{0,0,C2332},{0,C2332,0}},
{{C2233,0,0},{0,C2233,0},{0,0,C2222}}}}
```

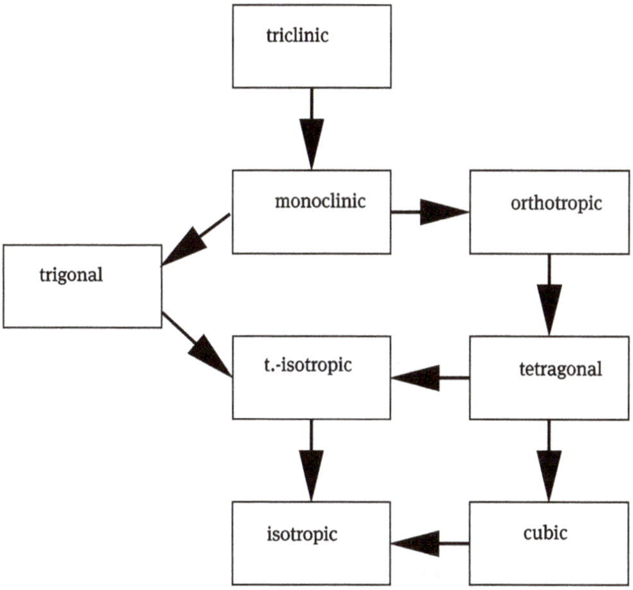

Figure 13.5: Subset relations of the symmetry classes of elasticity. The arrows point from the supergroup to the subgroup when referring to the set of possible stiffness tetrads with the corresponding symmetry. For example, one can see that every transversal isotropic material is as well trigonal and that the intersection between transversal isotropic and cubic materials are the isotropic materials. The converse holds not for the groups $\mathcal{G}$ of orthogonal tensors.

13.4 From volume averages to orientation averages

The Voigt mean of the stiffness is

$$\overline{\mathbb{C}} = \frac{1}{V_\Omega} \int_\Omega \mathbb{C}(\mathbf{x})\, \mathrm{d}v. \tag{13.73}$$

We can write the location-dependent stiffness in a crystal $\mathbb{C}(\mathbf{x})$ as a location-dependent rotation of a reference stiffness $\mathbb{C}_\#$,

Listing 13.2: Mathematica noebook (https://gitlab.com/gluegerainer/listings-homogenization-methods/-/blob/main/Listing-28_en.nb) for showing the rank reduction up to the *n*-fold symmetry for tensors of order *n*. For a tensor of order *n*, constraints between the components appear up to period $2\pi/n$.

```
Remove["Global`*"]
order = 2; (* tensorial order *)
CC = Array[C, ConstantArray[3, order]]; (* create symbolic tensor *)
Q = RotationMatrix[omega, {0, 0, 1}] (* create rotation matrix *);
(* define Rayleigh product for arbitrary order tensors *)
Rayleigh[rotmat_, arg_] := Module[
  {depth, tmp},
  depth = Depth[arg] - 2;
  tmp = arg;
  For[i = 1, i <= depth, i++,
   tmp = Transpose[tmp, 1 <-> i];
   tmp = rotmat.tmp;
   tmp = Transpose[tmp, 1 <-> i];
   ];
  tmp
  ]
(* extract linear system matrix and get its rank *)
matrix = CoefficientArrays[{Flatten[Rayleigh[Q, CC] - CC]},
    Flatten[CC]][[2, 1]];
MatrixForm[matrix]
rank = MatrixRank[matrix];
Print["Dimension of the tensor space: ", 3^order]
Print["Number of independent equations due to symmetry requirements: ", rank]
Print["Number of independent components: ", 3^order - rank]
erg = Solve[Flatten[Rayleigh[Q, CC] == CC], Flatten[CC]] // FullSimplify
CC /. erg[[1]] // MatrixForm
Print["For these angles there is a rank decrease: "]
Map[Solve[#1 == 0, omega] &, Eigenvalues[matrix]] // DeleteDuplicates
```

$$\overline{\mathbb{C}} = \frac{1}{V_\Omega} \int_\Omega \mathbf{Q}(\mathbf{x}) * \mathbb{C}_\# \, dv. \tag{13.74}$$

We find one orientation at each location, but the same orientation can appear at different locations. The function $\mathbf{Q}(\mathbf{x})$ is not invertible. Luckily, when calculating the volume average, we do not need the exact locations but only the volume fractions. This is reflected by the fact that $\mathbf{x}$ appears only indirectly via $\mathbf{Q}(\mathbf{x})$ in the integral in equation (13.74).

We introduce the probability density $d(\mathbf{Q})$ in the orientation space to convert the real space volume integral into an integral over orientation space. Let $\Delta SO(3)$ be a region in SO(3) with the volume fraction $g_{\Delta \mathrm{SO}(3)}$ and $\Delta\Omega$ be a region in real space with the sample volume $v_{\Delta\Omega}$. Then the quotient $d = g_{\Delta \mathrm{SO}(3)}/v_{\Delta\Omega}$ quantifies how big the real space volume in relation to orientation space volume needs to be for the probability of a random point in real space to lie in $\Delta\Omega$ being equal to the probability of a random point in orientation space to lie in $\Delta SO(3)$. The exact shape and location of $\Delta\Omega$ is not relevant, just its size, which is why the dependence on $\mathbf{x}$ vanishes.

However, we are interested in the position of $\Delta\mathbf{Q}$ in orientation space. The locations in SO(3) are contained via $\mathbf{Q} * \cdots$ in the integral, $\mathrm{d}v_{\mathbf{Q}}$ is the surrounding differential volume element in SO(3). Hence, we write

$$\frac{\mathrm{d}v}{V_\Omega} = d(\mathbf{Q})\,\mathrm{d}v_{\mathbf{Q}}, \tag{13.75}$$

with the probability density $d(\mathbf{Q})$. We replace the integral over the sample space with an integral over the entire orientation space,

$$\overline{\mathbb{C}} = \int\limits_{\mathrm{SO}(3)} d(\mathbf{Q})\mathbf{Q} * \mathbb{C}_{\#}\,\mathrm{d}v_{\mathbf{Q}}. \tag{13.76}$$

The integral over a region $\mathrm{A} \subset \mathrm{SO}(3)$

$$\mathrm{A} = [\phi_1 \ldots \phi_1 + \Delta\phi_1, \Phi \cdots \Phi + \Delta\Phi, \phi_2 \ldots \phi_2 + \Delta\phi_2] \tag{13.77}$$

in orientation space gives the probability that the orientation of a randomly selected sample point $\mathbf{x}$ lies in the region A:

$$p(\mathbf{Q}(\mathbf{x}) \in \mathrm{A}) = \frac{1}{8\pi^2} \int\limits_{\phi_2}^{\phi_2+\Delta\phi_2} \int\limits_{\Phi}^{\Phi+\Delta\Phi} \int\limits_{\phi_1}^{\phi_1+\Delta\phi_1} \sin\Phi\, g(\phi_1, \Phi, \phi_2)\,\mathrm{d}\phi_1\,\mathrm{d}\Phi\,\mathrm{d}\phi_2. \tag{13.78}$$

Here, the Euler angle parameterization has been chosen. $p(\mathbf{Q}(\mathbf{x}))$ is the probability distribution.

Let us consider the example of a rolled sheet in Figure 13.6. During the rolling, the crystal orientations align with respect to the rolling direction, such that orientations with ±45° w. r. t. the rolling direction are more frequent.

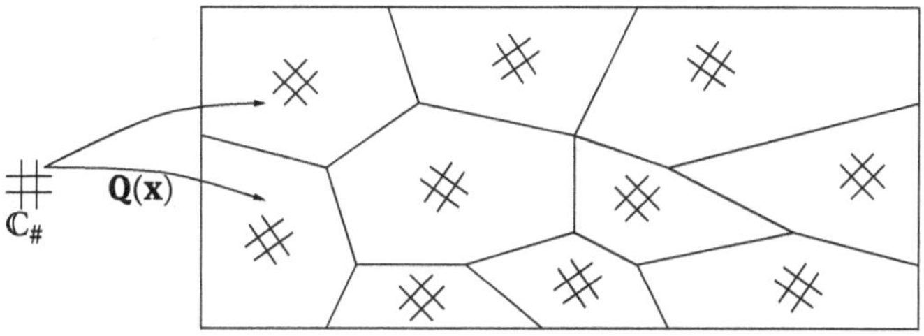

Figure 13.6: Sketch of a grain structure in a rolled sheet. The elongation of the grains in the rolling direction (horizontal) outlined. The orientation is outlined by the #-symbol. Orientations that deviate ≈45° from the rolling direction dominate.

13.4.1 Discrete orientation distributions

For n discrete orientations, the orientation distribution is

$$g = \sum_{i=1...n} g_i \delta(\mathbf{Q} - \mathbf{Q}_i), \tag{13.79}$$

with the Dirac distribution $\delta(\mathbf{Q})$. Upon integration, this becomes

$$\overline{\mathbb{C}} = \sum_{i=1...n} g_i \mathbf{Q}_i * \mathbb{C}_{\#}. \tag{13.80}$$

The g_i are the volume fractions $V_{\mathbf{Q}_i}/V_\Omega$. We could measure the 10 grains in Figure 13.6 and relate them to the overall sample size.

13.4.2 Continuous orientation distributions

Continuous orientation distributions are specified with the aid of the probability distribution $d(\mathbf{Q})$. Its properties are inherited from the properties of the ingredients g_i and v_i:

$$\text{Normality:} \quad \int_{SO(3)} d(\mathbf{Q})\, \mathrm{d}v_{\mathbf{Q}} = 1 \tag{13.81}$$

$$\text{Positivity:} \quad d(\mathbf{Q}) > 0 \quad \forall \mathbf{Q} \in SO(3). \tag{13.82}$$

In our example, $d(\mathbf{Q})$ could be a Gaussian bell-shaped distribution centered around the rotation ±45° w. r. t. the rolling direction. A normal distribution on the unit sphere in $\mathbb{R}_n$ is the von Mises[6]–Fisher[7] distribution; see Section 5.2.1 in Morawiec (2004). On the unit sphere in $\mathbb{R}_3$, the distribution is

$$p(\mathbf{x}) = \frac{\kappa}{2\pi(\mathrm{e}^{\kappa} - \mathrm{e}^{-\kappa})} \mathrm{e}^{\kappa \mathbf{x}\cdot\mathbf{d}}, \tag{13.83}$$

with $\|\mathbf{x}\| = \|\mathbf{d}\| = 1$, where $\mathbf{d}$ is the bell distribution center and κ a concentration parameter. It is depicted in Figure 13.7 for a small and a large concentration.

To get from the distribution on the unit sphere in $\mathbb{R}_3$ to a distribution in orientation space, one can use three interdependent Mises–Fisher distributions for a reoriented basis, or make use of the correspondence between spherical coordinates and two Euler angles, as discussed in the next section.

---

6 Richard von Mises, 1883–1953 (https://en.wikipedia.org/wiki/Richard_von_Mises).

7 Sir Ronald Aylmer Fisher, 1890–1962 (https://en.wikipedia.org/wiki/Ronald_Aylmer_Fisher).

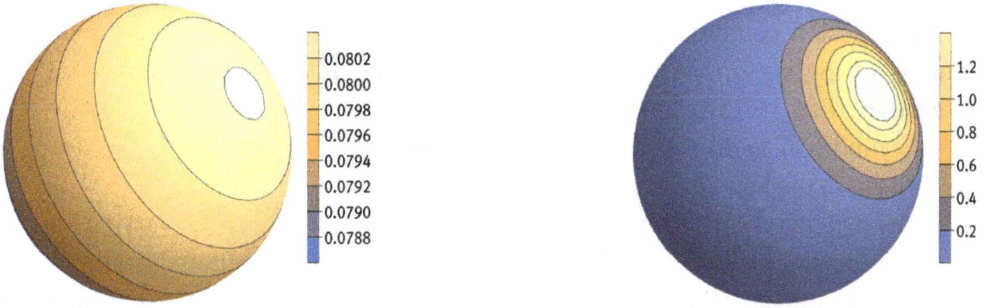

Figure 13.7: Von Mises–Fisher-distribution with $\kappa = 0.01$ (left) and $\kappa = 10$ (right). The low concentration leads to an almost isotropic orientation distribution with $p \approx 1/4/\pi \approx 0.0796$.

13.5 Integration over SO(3)

We will integrate orientation distributions over SO(3). To do so, we need a handle on the volume of SO(3). Until now, we have written

$$\int_{SO(3)} \mathrm{d}v_{\mathbf{Q}} \tag{13.84}$$

abstractly, without specifying $\mathrm{d}v_{\mathbf{Q}}$. For concrete calculations, we need to express $\mathrm{d}\mathbf{Q}$ and $\mathrm{d}v_{\mathbf{Q}}$ for the parameterization of choice. This is harder than for real space volume integrals, where we can always resort to a Cartesian basis.

Curvilinear coordinates in $\mathbb{R}_3$

In $\mathbb{R}_3$, we can always use $\mathrm{d}V = \mathrm{d}x\,\mathrm{d}y\,\mathrm{d}z$ and the coordinate conversion to obtain $\mathrm{d}V$ w. r. t. other coordinates. For example, to parameterize $\mathbb{R}_3$ with cylindrical coordinates, we use the coordinate conversion

$$x = r\cos\phi \tag{13.85}$$

$$y = r\sin\phi \tag{13.86}$$

$$z = z. \tag{13.87}$$

First, one determines the basis $\mathbf{g}_i$ that is location dependent and tangential to the coordinate lines by partial derivatives w. r. t. the coordinates:

$$\mathbf{x} = x\mathbf{e}_x + y\mathbf{e}_y + z\mathbf{e}_z \tag{13.88}$$

$$\mathbf{g}_r = \frac{\partial\mathbf{x}}{\partial r} = \cos\phi\mathbf{e}_x + \sin\phi\mathbf{e}_y \tag{13.89}$$

$$\mathbf{g}_\phi = \frac{\partial\mathbf{x}}{\partial\phi} = -r\sin\phi\mathbf{e}_x + r\cos\phi\mathbf{e}_y \tag{13.90}$$

$$\mathbf{g}_z = \frac{\partial\mathbf{x}}{\partial z} = \mathbf{e}_z \tag{13.91}$$

In Cartesian coordinates, the coordinates of a location vector equal the components of the vector w. r. t. the basis $\mathbf{e}_i$. The differential volume element $\mathrm{d}V = [\mathrm{d}x\mathbf{e}_x, \mathrm{d}y\mathbf{e}_y, \mathrm{d}z\mathbf{e}_z]$ is the same everywhere.

In cylindrical coordinates, the location is given by coordinates r, ϕ, z, but w. r. t. which basis is the location vector denoted.

The tangential basis is the differential of the location vector when varying the coordinates (partial derivatives). For ϕ, it depends linearly on the radius. Therefore, the volume elements grow linearly with r. This dependence is usually removed by normalizing the tangential basis. To write down the location vector, one needs to choose a vector basis. A canonical choice is a basis at the said location. This is $\mathbf{x} = r\mathbf{g}_r(\phi) + z\mathbf{g}_z$. Note that the coordinates and the location vector components are different and depend on the basis. A differential volume is the triple product of $\mathrm{d}r\,\mathbf{g}_r$, $\mathrm{d}\phi\,\mathbf{g}_\phi$ and $\mathrm{d}z\,\mathbf{g}_z$; see Figure 13.8. It is linear in the differentials such that

$$\mathrm{d}V = [\mathbf{g}_r, \mathbf{g}_\phi, \mathbf{g}_z]\,\mathrm{d}r\,\mathrm{d}\phi\,\mathrm{d}z \tag{13.92}$$

holds. The triple product $[\mathbf{g}_r, \mathbf{g}_\phi, \mathbf{g}_z]$ is therefore the determinant of the Jacobian matrix,

$$[\mathbf{g}_r, \mathbf{g}_\phi, \mathbf{g}_z] = \det \underbrace{\begin{bmatrix} \cos\phi & \sin\phi & 0 \\ -r\sin\phi & r\cos\phi & 0 \\ 0 & 0 & 1 \end{bmatrix}}_{J_{ij}} . \tag{13.93}$$

Moreover, there is the metric. It contains the length of the tangential base vectors as well as the angles between them:

$$G_{ij} = \begin{bmatrix} \mathbf{g}_r\cdot\mathbf{g}_r & \mathbf{g}_r\cdot\mathbf{g}_\phi & \mathbf{g}_r\cdot\mathbf{g}_z \\ \mathbf{g}_\phi\cdot\mathbf{g}_r & \mathbf{g}_\phi\cdot\mathbf{g}_\phi & \mathbf{g}_\phi\cdot\mathbf{g}_z \\ \mathbf{g}_z\cdot\mathbf{g}_r & \mathbf{g}_z\cdot\mathbf{g}_\phi & \mathbf{g}_z\cdot\mathbf{g}_z \end{bmatrix} = \begin{bmatrix} 1 & 0 & 0 \\ 0 & r^2 & 0 \\ 0 & 0 & 1 \end{bmatrix} . \tag{13.94}$$

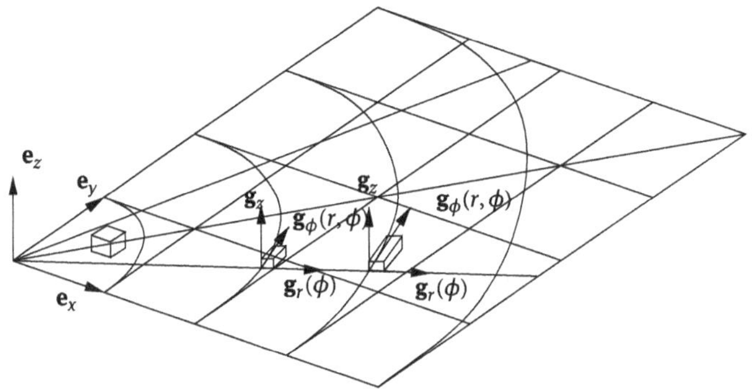

Figure 13.8: For Cartesian coordinates, the basis $\mathbf{e}_i$ is location independent. For curvilinear coordinates, the basis $\mathbf{g}_i$ tangential to the coordinate lines depends on the location.

One can see that the coordinate lines intersect under 90°, and the tangential basis is normalized except for $\mathbf{g}_\phi$. Further, and more generally,

$$G_{ij} = J_{ik}J^T_{kj} \tag{13.95}$$

holds. With the determinant rule, we can therefore identify

$$\det(J_{ij}) = \sqrt{\det(G_{ij})}. \tag{13.96}$$

Curvilinear coordinates in SO(3)

When one tries to do the same for SO(3), one immediately encounters problems since there is no Cartesian parameterization w. r. t. which one can work, i. e., in a sense, the starting point is missing. A Cartesian parameterization cannot exist due to the doubly connected topology of SO(3). At best, we can try to convert between different non-Cartesian coordinates.

The cause and solution to this complication lie in the fact that SO(3) is a Lie[8] group. The Lie group has a corresponding Lie algebra. The deviation between infinitesimal rotations is not a difference, as is the case for location vectors in $\mathbb{R}_3$, but it is a product, like the differential generators in Section 13.1. The differential changes of $dv_{\mathbf{Q}(c_1,c_2,c_3)}$ that result upon varying the coordinates c_i are not elements of SO(3), but antisymmetric tensors.

The latter can be reduced to axial vectors in $\mathbb{R}_3$. With the cross product, these form the Lie algebra. Moreover, the scalar product can be calculated between them, such that we can determine the metric just as in equation (13.94) also for coordinates in SO(3). This can be used to calculate the determinant of J_{ij} according to equation (13.96).

Here, we use the Euler angle parameterization. Deviations between rotations are obtained in SO(3) as products with inverse rotations. For example,

$$\mathbf{W}_{\phi_1} = \left.\frac{\partial \mathbf{Q}(\phi_1, \Phi, \phi_2)^T \mathbf{Q}(\phi_1 + \Delta\phi_1, \Phi, \phi_2)}{\partial \Delta\phi_1}\right|_{\Delta\phi_1=0} \tag{13.97}$$

$$= \mathbf{Q}(\phi_1, \Phi, \phi_2)^T \frac{\partial \mathbf{Q}(\phi_1, \Phi, \phi_2)}{\partial \phi_1} = \begin{bmatrix} 0 & -1 & 0 \\ 1 & 0 & 0 \\ 0 & 0 & 0 \end{bmatrix}. \tag{13.98}$$

Analogously, we obtain

$$\mathbf{W}_\Phi = \mathbf{Q}(\phi_1, \Phi, \phi_2)^T \frac{\partial \mathbf{Q}(\phi_1, \Phi, \phi_2)}{\partial \Phi} = \begin{bmatrix} 0 & 0 & -\sin\phi_1 \\ 0 & 0 & -\cos\phi_1 \\ \sin\phi_1 & \cos\phi_1 & 0 \end{bmatrix} \tag{13.99}$$

8 Sophus Lie, 1842–1899 (https://en.wikipedia.org/wiki/Sophus_Lie).

$$\mathbf{W}_{\phi_2} = \mathbf{Q}(\phi_1, \Phi, \phi_2)^T \frac{\partial \mathbf{Q}(\phi_1, \Phi, \phi_2)}{\partial \phi_2} = \begin{bmatrix} 0 & -\cos\Phi & \cos\phi_1 \sin\Phi \\ \cos\Phi & 0 & -\sin\phi_1 \sin\Phi \\ -\cos\phi_1 \sin\Phi & \sin\phi_1 \sin\Phi & 0 \end{bmatrix}. \tag{13.100}$$

We now have the counterpart to the tangential basis $\mathbf{g}_i$ from the example with cylindrical coordinates. The scalar products can be calculated in various ways: We can determine $\mathbf{W}_i : \mathbf{W}_j$ directly, or we can first determine the axial vectors $\mathbf{w}_i = -\frac{1}{2}\overset{\langle 3 \rangle}{\boldsymbol{\varepsilon}} : \mathbf{W}$ (equation (13.9)) and then determine the scalar product between these. Depending on the definition, the signs and the magnitudes may vary. This is nevertheless unimportant since we normalize the volume of SO(3) anyway. We therefore stick to the scalar products between the $\mathbf{W}_i$,

$$G_{ij}^{\mathrm{SO(3)_{Euler}}} = \begin{bmatrix} \mathbf{W}_{\phi_1} : \mathbf{W}_{\phi_1} & \mathbf{W}_{\phi_1} : \mathbf{W}_{\Phi} & \mathbf{W}_{\phi_1} : \mathbf{W}_{\phi_2} \\ \mathbf{W}_{\Phi} : \mathbf{W}_{\phi_1} & \mathbf{W}_{\Phi} : \mathbf{W}_{\Phi} & \mathbf{W}_{\Phi} : \mathbf{W}_{\phi_2} \\ \mathbf{W}_{\phi_2} : \mathbf{W}_{\phi_1} & \mathbf{W}_{\phi_2} : \mathbf{W}_{\Phi} & \mathbf{W}_{\phi_2} : \mathbf{W}_{\phi_2} \end{bmatrix} = \begin{bmatrix} 2 & 0 & 2\cos\Phi \\ 0 & 2 & 0 \\ 2\cos\Phi & 0 & 2 \end{bmatrix}. \tag{13.101}$$

The determinant of the metric is then

$$\det(G_{ij}^{\mathrm{SO(3)_{Euler}}}) = 8(1 - \cos^2\Phi) = 8\sin^2\Phi, \tag{13.102}$$

by which the determinant of the Jacobian matrix is after equation (13.96),

$$J = \sqrt{G_{ij}^{\mathrm{SO(3)_{Euler}}}} = \sqrt{8}\sin\Phi. \tag{13.103}$$

Equivalence of the Jacobian determinant of SO(3) and spherical coordinates

This is a somewhat surprising result. The Jacobian determinant for the parameterization of SO(3) with Euler angles in the $z-x-z$ convention is, upon amending the term $r^2/\sqrt{8}$, identical to the Jacobian determinant when $\mathbb{R}_3$ is parameterized with spherical coordinates with the latitude measured from the pole; see equation (12.46) or equation (12.185). We can therefore transfer methods that have been developed for spherical coordinates to the Euler angle parameterization of SO(3), wherein SO(3) all three coordinates have finite, periodically repeating intervals, while in $\mathbb{R}_3$ the radius goes from 0 to ∞. Especially for the two angular spherical coordinates, the well-known spherical harmonics are useful as basis functions for approximating smooth orientation distributions. Two pioneering works in this regard are the books of Wigner (1931)[9] and Bunge (1969).[10]

9 Eugene Paul Wigner, 1902–1995 (https://en.wikipedia.org/wiki/Eugene_Paul_Wigner).

10 Hans-Joachim Bunge, 1929–2004 (https://de.wikipedia.org/wiki/Hans-Joachim_Bunge).

Normalization of the volume of SO(3)

We can now integrate over the volume of SO(3) by using Euler angles,

$$V_{\mathrm{SO(3)}} = \int_0^{2\pi}\int_0^{\pi}\int_0^{2\pi} \sqrt{8}\sin\Phi\,\mathrm{d}\phi_1\,\mathrm{d}\Phi\,\mathrm{d}\phi_2 \tag{13.104}$$

$$= 16\sqrt{2}\pi^2 \tag{13.105}$$

To normalize the volume, we can expand the Jacobian determinant by the factor $(16\sqrt{2}\pi^2)^{-1}$, which finally gives

$$\int_{\mathrm{SO(3)}} \cdot\,\mathrm{d}V = \frac{1}{8\pi^2}\int_0^{2\pi}\int_0^{\pi}\int_0^{2\pi} \cdot\sin\Phi\,\mathrm{d}\phi_1\,\mathrm{d}\Phi\,\mathrm{d}\phi_2 \tag{13.106}$$

for integrating orientation distributions over SO(3). The following Mathematica Listing 13.3 derives the Jacobian determinant.

Listing 13.3: Mathematica notebook (https://gitlab.com/gluegerainer/listings-homogenization-methods/-/blob/main/Listing-29_en.nb) to calculate the Jacobian determinant needed for integration over SO(3) when parameterized by Euler angles.

```
Remove["Global`*"];
W={0,0,0}; (* Empty list to collect the tangential basis *)
Q[args_]:=(* Q parameterized by Euler angles in the z-x-z-convention *)
RotationMatrix[args[[3]],{0,0,1}].
RotationMatrix[args[[2]],{1,0,0}].
RotationMatrix[args[[1]],{0,0,1}];
QT = Transpose[Q[{phi1, phi, phi2}]];
Q = Q[{phi1, phi, phi2}];
W[[1]] = QT.D[Q, phi1] // FullSimplify;
W[[2]] = QT.D[Q, phi] // FullSimplify;
W[[3]] = QT.D[Q, phi2] // FullSimplify;
Metrik=Table[FullSimplify[Tr[W[[i]].Transpose[W[[j]]]]],{i,1,3},{j,1,3}];
Print["Jacobian determinant for the normalized volume in Euler angles:"]
Jacobi=Sqrt[Simplify[Det[Metrik]]]
Print["Volume of SO(3) in Euler angles:"]
Integrate[Sqrt[Det[Metrik]],{phi1,0,2 Pi},{phi2,0,2 Pi},{phi,0,Pi}]
```

To do the same for the Rodriguez parameterization, we need to replace the defining function for the rotation **Q**, see Listing 13.4, by which we obtain the metric and Jacobian determinant contained in the below integral,

$$G^R_{ij} = \frac{1}{(1+w_1^2+w_2^2+w_3^2)^2}\begin{bmatrix} 8(w_2^2+w_3^2+1) & -8w_1w_2 & -8w_1w_3 \\ -8w_1w_2 & 8(w_1^2+w_3^2+1) & -8w_2w_3 \\ -8w_1w_3 & -8w_2w_3 & 8(w_1^2+w_2^2+1) \end{bmatrix} \tag{13.107}$$

$$\int_{\mathrm{SO(3)}} \cdot\,\mathrm{d}V = \frac{1}{\pi^2}\int_{-\infty}^{\infty}\int_{-\infty}^{\infty}\int_{-\infty}^{\infty} \cdot\sqrt{\frac{1}{(1+w_1^2+w_2^2+w_3^2)^4}}\,\mathrm{d}w_1\,\mathrm{d}w_2\,\mathrm{d}w_3 \tag{13.108}$$

Listing 13.4: Modification (https://gitlab.com/gluegerainer/listings-homogenization-methods/-/blob/main/Listing-29_en.nb) of Listing 13.3 for the Rodrigues parameterization.

```
Q[args_]:=(* This is the Rodrigues parameterization of Q *)
(WR=-LeviCivitaTensor[3].args;
id=IdentityMatrix[3];
Inverse[id+WR].(id-WR))
```

For the axis-angle parameterization, this calculation escalates due to nested trigonometric and exponential functions.

13.6 Graphical representation of orientation distributions

The depiction of orientation distributions is challenging due to the topology of SO(3) and its dimensionality 3.

One may think of plotting an intensity in Cartesian coordinates for points in the axis-angle sphere with radius π. In such a plot, rotations by small angles are concentrated near the origin, and rotations by 180° are spread out near the surface. Hence, a homogeneous distribution in SO(3) does not have a homogeneous depiction in the axis-angle sphere.

Typical representations are pole figures, which are stereographic projections. The endpoints of a rotated basis are projected along rays from a pole into the equatorial plane, as sketched in Figure 13.9. Such images are, in fact, measured as diffraction patterns, which is why this representation is useful for a direct comparison to measurements. Unfortunately, this projection is not area preserving, such that homogeneous distributions on the sphere appear inhomogeneous in the pole figure.

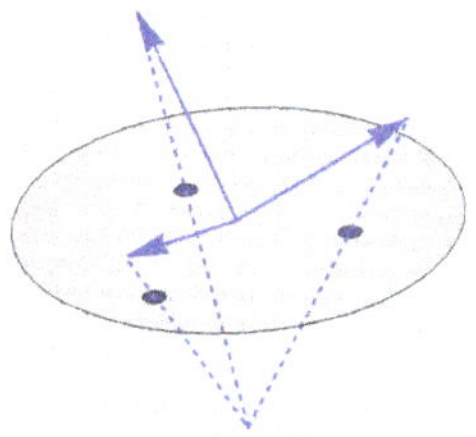

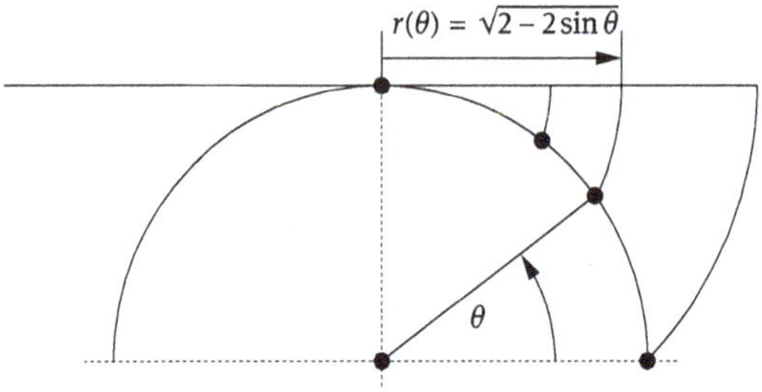

Figure 13.9: Left: Stereographic projection of the endpoints of a rotated basis into the equatorial plane. Right: Area preserving azimuthal projection of circles of latitudes.

For visualizing the anisotropy, an area-preserving azimuthal or Lambert[11] projection is more useful. It is obtained by setting a differentially small area element from the

11 Johann Heinrich Lambert, 1728–1777 (https://en.wikipedia.org/wiki/Johann_Heinrich_Lambert).

unit sphere $\mathrm{d}A = \cos\theta\,\mathrm{d}\theta\,\mathrm{d}\phi$ (with the angle θ being zero at the equator) equal with a differentially small area element in the equatorial plane in polar coordinates $\mathrm{d}A = r\,\mathrm{d}r\,\mathrm{d}\phi$. After canceling out $\mathrm{d}\phi$, one finds

$$\cos\theta\,\mathrm{d}\theta = r\,\mathrm{d}r \tag{13.109}$$

$$r'(\theta) = \frac{\cos\theta}{r}. \tag{13.110}$$

This differential equation can be solved for $r(\theta)$ with the boundary condition $r(\pi/2) = 0$, i. e., that the north pole lies in the center at $r = 0$ in the polar projection plane. The result is

$$r(\theta) = \sqrt{2 - 2\sin\theta}. \tag{13.111}$$

Mapping the spherical coordinate grid with this projection gives the Schmidt[12] net.[13]

Another way to represent orientation distributions graphically is to use intensity plots of sections of the Euler angle space. These are less easily interpreted, and the poles at $\Phi = 0$ and $\Phi = \pi$ appear stretched and overrepresented. It may be because the conversion of intensities from pole figures to orientations was first done by Roe (1965), Bunge (1965) in Euler angles, which is why this is an established method; see, e. g., Böhlke (2005).

13.7 Generation of discrete isotropic distributions in SO(3)

In light of the above, one may wonder how isotropic orientation distributions can be generated. It is obvious that uniform, random distributions in any parameterization of SO(3) are inadequate due to the curvilinear nature of any such parameterization. In any parameterization, there are regions with varying coordinate line densities, similar to the concentration of meridians at the earth's poles.

To compensate this effect, the random variable needs to be weighed with the Jacobian determinant, which scales with the size of a differentially small volume element spanned by the coordinate lines. The smaller the Jacobian determinant, the smaller the probability that the differential volume element contains a randomly chosen orientation from an isotropic distribution.

Let us examine the Euler angles. We have seen that $J = \sin(\Phi)/8/\pi^2 = p_{\phi_1} p_{\phi_2} p_{\Phi}$. J does not depend on $\phi_{1,2}$, which is why a homogeneous orientation distribution in SO(3) implies a homogeneous distribution of the angles $\phi_{1,2}$. We can hence presume

12 Walter Schmidt, 1885–1945 (https://de.wikipedia.org/wiki/Walter_Schmidt_(Geologe)).

13 Schmidt net (https://en.wikipedia.org/wiki/Schmidt_net).

homogeneous (or constant or uniform) probability densities for $p_{\phi_{1,2}} = 1/2/\pi$ in the interval $0 < \phi_{1,2} < 2\pi$. It remains

$$p_{\Phi} = \frac{\sin\Phi}{2} \tag{13.112}$$

for $0 < \Phi < \pi$. The integral over $0 < \Phi < \pi$ gives the value 1; see Figure 13.10. The integral from 0 to $\underline{\Phi}$ is the cumulative probability density

$$c_{\underline{\Phi}} = \int_0^{\underline{\Phi}} p_{\Phi}\,\mathrm{d}\Phi = \frac{1}{2}(1 - \cos\underline{\Phi}). \tag{13.113}$$

It gives the probability of a random angle that is distributed with p_{Φ} lies in the interval $0\ldots\underline{\Phi}$. A homogeneous (or constant or uniform) probability density has, therefore, a linear cumulative probability density. We can now ask the reverse question: Which random, nonhomogeneous distribution gives after multiplication by p_{Φ} upon integration a linear cumulative probability density? We just need to equate $c_{\underline{\Phi}}$ with a linear cumulative probability density mx and solve for $\underline{\Phi}$,

$$mx = \frac{1}{2}(1 - \cos\underline{\Phi}). \tag{13.114}$$

We chose for x the interval $0 \le x \le 1$, such that $m = 1$. This gives a conversion rule for random values x from a homogeneous distribution over an interval $0\ldots1$ into values $\underline{\Phi}$ of an inhomogeneous probability density in the interval $0\ldots\pi$ with the probability density $p_{\underline{\Phi}}$. We obtain

$$\underline{\Phi} = \arccos(1 - 2x). \tag{13.115}$$

The concentration of orientations when (wrongly) using homogeneous distributions is depicted in Figure 13.11, which has been generated by the script in Listing 13.5. Brannon (2018), Chapter 17, is very detailed and generously illustrated regarding the generation of discrete isotropic distributions.

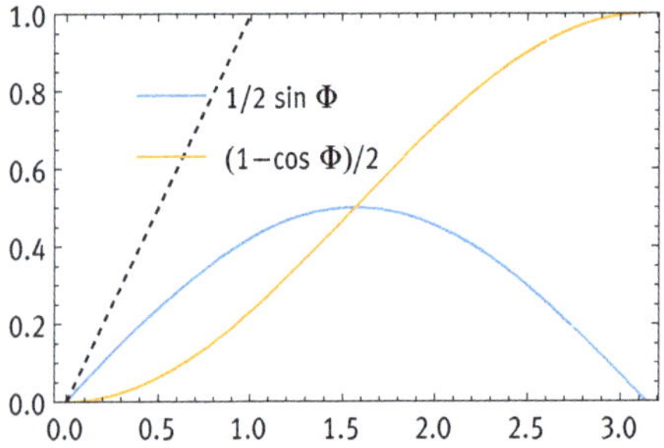

Figure 13.10: Probability density function p_{Φ} (blue) and cumulative density function $c_{\underline{\Phi}}$ (yellow).

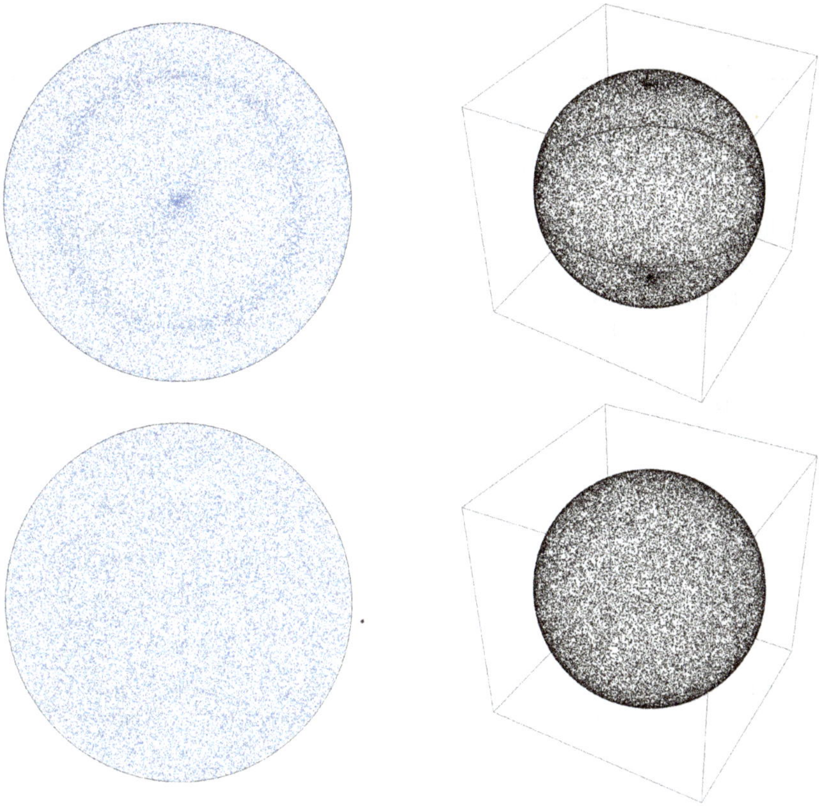

Figure 13.11: Depiction of discrete orientation distributions in the area-preserving Lambert-projection (left) and on a sphere, above with and below without accounting for the inhomogeneity of the random distribution of the Euler angle Φ; see Listing 13.5.

Listing 13.5: Mathematica notebook (https://gitlab.com/gluegerainer/listings-homogenization-methods/-/blob/main/Listing-31_en.nb) for generating homogeneous discrete distributions in SO(3).

```
Remove["Global`*"]
cartesiantospherical[{x1_, x2_, x3_}] = {ArcCos[x3], ArcTan[x1, x2]};  (* Determine latitude and
      longitude *)
sphericalanglestolambert[{theta_, phi_}] = {phi, Sqrt[2 (1 - Cos[theta])]}; (* Area preserving
      projection *)
Q =  RotationMatrix[phi2, {0, 0, 1}]. (* Q from Euler angles *)
     RotationMatrix[phi, {1, 0, 0}].
     RotationMatrix[phi1, {0, 0, 1}] // FullSimplify;
oris = Flatten[Table[ (* Column vectors give points on unit sphere *)
   phi2 = 2 Pi RandomReal[];
   phi1 = 2 Pi RandomReal[];
   phi = ArcCos[1 - 2 RandomReal[]];  (* For isotropic orientation distributions *)
   (* Uncommenting the next line invokes a homogeneous distribution of phi, *)
   (* resulting in an anisotropic orientation distribution *)
   (* phi=Pi RandomReal[]; *)
   Q, {i, 10000}], 1];
points = Map[sphericalanglestolambert, Map[cartesiantospherical, oris]];
```

```
ListPolarPlot[points] (* Graphical output *)
Graphics3D[{PointSize[0.001], Point[oris]}]
```

13.7.1 Isotropic orientation distributions for arbitrary dimensions

A nifty method to obtain a random orientation in dimension d is to generate an instance G_{ij} of the Gaussian orthogonal matrix distribution of size $d \times d$ and calculate the singular value decomposition $G_{ij} = U_{im}\Sigma_{mn}V^T_{nj}$. The matrices U_{im} and V_{jn} are random, uniformly distributed orthogonal matrices over $\mathbb{R}_d$.

The properties of G_{ij} are

- symmetric, $G_{ij} = G_{ji}$
- the off-diagonal entries G_{ij}, $i \neq j$ are normal distributed around 0 with the variance v
- the diagonal entries G_{ii} are normally distributed around 0 with the variance $2v$

The variance v affects only Σ_{mn} and can be chosen freely. Listing 13.6 contains an implementation and a demonstration for $\mathbb{R}_2$ and $\mathbb{R}_3$, the generated distributions are depicted in Figure 13.12.

Listing 13.6: Mathematica notebook (https://gitlab.com/gluegerainer/listings-homogenization-methods/-/blob/main/Listing-32_en.nb) for generating homogeneous discrete distributions in SO(d) and viewing them for $d = 2$ on a circle and $d = 3$ on a sphere.

```
Remove["Global`*"]
dim = 2; (* Dimension > 1, graphics only for 2 and 3 *)
n = 100; (* Sample size *)
oris = Map[SingularValueDecomposition,
          RandomVariate[GaussianOrthogonalMatrixDistribution[dim], n]][[;;,1]];
(* Draw rotated base vectors *)
If[dim == 2, Graphics[Point[Flatten[oris, 1]]]]
If[dim == 3, Graphics3D[Point[Flatten[oris, 1]]]]
```

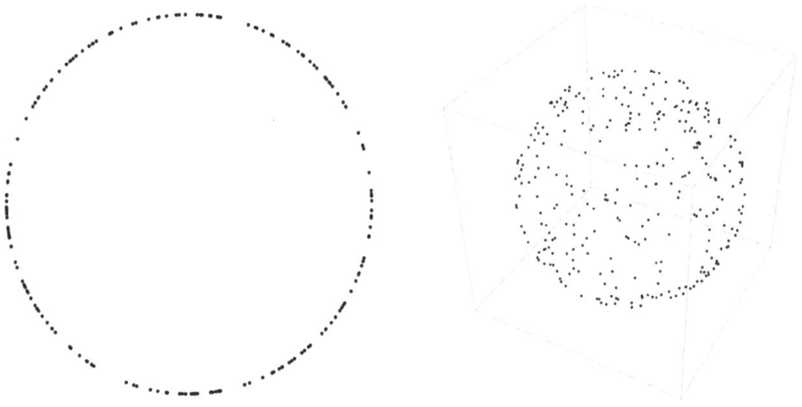

Figure 13.12: 100 random orientations in $\mathbb{R}_2$ and $\mathbb{R}_3$ as generated by Listing 13.6.

13.7.2 Orientation mean value

The mean value of orthogonal tensors $\overline{\mathbf{Q}} = \sum v_i \mathbf{Q}_i$ over a discrete distribution in SO(3) is itself not an orthogonal tensor due to the fact that SO(3) has a multiplicative structure. Nevertheless, it can serve as an indicator for symmetric and isotropic distributions, similar to the vanishing integral $\int \mathbf{n}\, \mathrm{d}A = \mathbf{o}$ over all normal vectors over a closed surface $\overline{\mathbf{Q}} = \mathbf{O}$ in the case of symmetric and isotropic distributions.

13.8 Integration over isotropic distributions in SO(3)

In case of isotropic distributions, we have $d = 1/V_{\mathrm{SO}(3)}$ if the Jacobian determinant has not been normalized already with $V_{\mathrm{SO}(3)}$. Here, we summarize this normalization with the Jacobian determinant. Hence, here we have $d = 1$. The key point is that d is homogeneous (or uniform or constant), i. e., every orientation has the same probability. With the rotation of a tensor of higher order (equation (13.57)), for a fourth-order tensor, we have to evaluate the expression

$$\overline{\mathbb{C}} = \underbrace{\int_0^{2\pi}\int_0^{\pi}\int_0^{2\pi} \frac{\sin\Phi}{8\pi^2} Q_{im}^{\phi_1,\Phi,\phi_2} Q_{jn}^{\phi_1,\Phi,\phi_2} Q_{ko}^{\phi_1,\Phi,\phi_2} Q_{lp}^{\phi_1,\Phi,\phi_2}\, \mathrm{d}\phi_1\, \mathrm{d}\Phi\, \mathrm{d}\phi_2}_{\overset{\langle 8\rangle}{\mathbb{P}} :::: \mathbf{e}_i\otimes\mathbf{e}_j\otimes\mathbf{e}_k\otimes\mathbf{e}_l\otimes\mathbf{e}_m\otimes\mathbf{e}_n\otimes\mathbf{e}_o\otimes\mathbf{e}_p}\; C_{\#mnop}\mathbf{e}_i \otimes \mathbf{e}_j \otimes \mathbf{e}_k \otimes \mathbf{e}_l \tag{13.116}$$

for integration over SO(3) in terms of Euler angles. The components $Q_{ij}^{\phi_1,\Phi,\phi_2}$ depend on ϕ_1, Φ and ϕ_2, but we can pull the constant reference stiffness $\mathbb{C}_\#$ out of the integral. The integral can then be considered as the component of an eighth-order tensor $\overset{\langle 8\rangle}{\mathbb{P}}$.

Evaluating these integrals is a tedious but doable task that is best left to a computer algebra system, exemplified in Listing 13.7. An advantage of the Euler angle parameterization is that only sine and cosine functions need to be integrated over multiples of π. Without using the symmetries, we have $3^8 = 6561$ index combinations of the indices i, j, k, l, m, n, o, p. With symmetries, we draw 4 out of 9 components of Q_{ij}, with placing back, and without taking into account the order of the drawn components. One finds that

$$\binom{9+4-1}{4} = 495 \tag{13.117}$$

different integrals result. In equation (13.116), one can already see that the application of an orientation average can be written as a linear mapping,

$$\overline{\mathbb{C}} = \overset{\langle 8\rangle}{\mathbb{P}} :: \mathbb{C}_\#. \tag{13.118}$$

We use the capital letter $\mathbb{P}$ due to the projector properties: They map the anisotropic part of $\mathbb{C}_\#$ into the fourth-order null tensor. We can use this property to significantly reduce the effort of finding $\mathbb{P}$.

Vectors

We determine exemplary the isotropic orientation average of the unit vector $\mathbf{e}_1$. Imagining a set of unit vectors homogeneously distributed on the unit sphere makes clear that they cancel out in a summation. Mathematically, this is obtained by integration over individual components of $\mathbf{Q}$ in terms of Euler angles (see equation (13.40)),

$$\frac{1}{8\pi^2}\int_0^{2\pi}\int_0^{\pi}\int_0^{2\pi} \sin\Phi\, Q_{ij}\mathbf{e}_i \otimes \mathbf{e}_j \,\mathrm{d}\phi_1\,\mathrm{d}\Phi\,\mathrm{d}\phi_2 \cdot \mathbf{e}_1, \tag{13.119}$$

with

$$Q_{ij} = \begin{bmatrix} \cos\phi_1\cos\phi_2 - \cos\Phi\sin\phi_1\sin\phi_2 & -\cos\phi_2\sin\phi_1 - \cos\Phi\cos\phi_1\sin\phi_2 & \sin\Phi\sin\phi_2 \\ \cos\Phi\cos\phi_2\sin\phi_1 + \cos\phi_1\sin\phi_2 & \cos\Phi\cos\phi_1\cos\phi_2 - \sin\phi_1\sin\phi_2 & -\cos\phi_2\sin\Phi \\ \sin\Phi\sin\phi_1 & \cos\phi_1\sin\Phi & \cos\Phi \end{bmatrix}. \tag{13.120}$$

The integrals over the intervals $0 \leq \phi_{1,2} < 2\pi$ are over full periods of sine and cosine functions and, therefore, zero, the integral over $0 \leq \Phi < \pi$ becomes with the factor $\sin\Phi$ of the Jacobian determinant an integral over a full period, such that all nine integrals vanish and the zero tensor maps all vectors into their isotropic parts.

Tensors of second order

For second-order tensors, we can see by $\mathbf{Q} * \mathbf{A} = \mathbf{Q}\mathbf{A}\mathbf{Q}^T = \mathbf{A}$ that only $\mathbf{A} = \alpha\mathbf{I}$ is invariant under arbitrary rotations. Hence, it remains only the dilatoric part

$$\mathbf{A}^\circ = \mathrm{tr}(\mathbf{A})/3\,\mathbf{I} = \underbrace{\frac{1}{3}\mathbf{I}\otimes\mathbf{I}}_{\mathbb{P}_{\mathrm{I}1}} : \mathbf{A} \tag{13.121}$$

upon isotropic orientation averaging a second-order tensor. The first isotropic projector $\mathbb{P}_{\mathrm{I}1}$ is obtained upon integration,

$$\mathbb{P}_{\mathrm{I}1} = \frac{1}{8\pi^2}\int_0^{2\pi}\int_0^{\pi}\int_0^{2\pi} \sin\Phi Q_{ik}Q_{jl}\,\mathrm{d}\phi_1\,\mathrm{d}\Phi\,\mathrm{d}\phi_2\, \mathbf{e}_i \otimes \mathbf{e}_j \otimes \mathbf{e}_k \otimes \mathbf{e}_l, \tag{13.122}$$

which is shown in Listing 13.7. However, it is obtained more easily by comparing coefficients in equation (13.121).

Listing 13.7: Mathematica notebook (https://gitlab.com/gluegerainer/listings-homogenization-methods/-/blob/main/Listing-33_en.nb) for integrating over SO(3) to obtain the first isotropic projector.

```
Remove["Global`*"];
(* Parameterization of Q by Euler angles *)
Q=RotationMatrix[phi2,{0,0,1}].RotationMatrix[phi,{1,0,0}].RotationMatrix[phi1,{0,0,1}];

Print["The isotropic orientation average of a unit vector and the corresponding projector:"]
Integrate[Q.{1,0,0}/8 /Pi^2Sin[phi],{phi2,0,2 Pi},{phi,0,Pi},{phi1,0,2Pi}]
Integrate[Q/8 /Pi^2Sin[phi],{phi2,0,2 Pi},{phi,0,Pi},{phi1,0,2Pi}]

Print["The isotropic orientation average of a second-order tensor and the corresponding
     projector:"]
A={{A11,A12,A13},{A12,A22,A23},{A13,A23,A33}};
Integrate[Q.A.Transpose[Q]/8 /Pi^2Sin[phi],{phi2,0,2 Pi},{phi,0,Pi},{phi1,0,2Pi}]
PI1=Integrate[Transpose[Outer[Times,Q,Q],2<->3]/8 /Pi^2Sin[phi],{phi2,0,2 Pi},{phi,0,Pi},{phi1
     ,0,2Pi}]
```

Tensors of third order

In the case of third-order tensors, one finds that only multiples of the permutation tensor are isotropic,

$$\mathbf{Q} * \left(\alpha \overset{\langle 3\rangle}{\boldsymbol{\varepsilon}}\right) = \alpha \overset{\langle 3\rangle}{\boldsymbol{\varepsilon}}, \tag{13.123}$$

which implies that the projector

$$\overset{\langle 6\rangle}{\mathbb{P}} = \frac{1}{6} \overset{\langle 3\rangle}{\boldsymbol{\varepsilon}} \otimes \overset{\langle 3\rangle}{\boldsymbol{\varepsilon}} \tag{13.124}$$

maps every third-order tensor into its isotropic part.

Stiffness tetrads

In the case of fourth-order tensors, we have several isotropic parts, but we are only interested in fourth-order tensors that have the principle and subsymmetries. We know that the spectral representation of isotropic stiffness tetrads is

$$\mathbb{C}_{\mathrm{Iso}} = 3K\mathbb{P}_{\mathrm{I1}} + 2G\mathbb{P}_{\mathrm{I2}}. \tag{13.125}$$

$\mathbf{P}_{\mathrm{I1,2}}$ are the eigenprojectors, $3K$ and $2G$ the eigenvalues. The projection of an arbitrary stiffness $\mathbb{C}$ onto this subspace is achieved by the eighth-order projector

$$\overset{\langle 8\rangle}{\mathbb{P}} = \mathbb{P}_{\mathrm{I1}} \otimes \mathbb{P}_{\mathrm{I1}} + \frac{1}{5}\mathbb{P}_{\mathrm{I2}} \otimes \mathbb{P}_{\mathrm{I2}}; \tag{13.126}$$

see Section 2.4 on the projectors $\mathbf{P}_{\mathrm{I1,2}}$.

13.9 Application of the orientation average

13.9.1 The heat conductivity of black phosphor

At room temperature, the heat conduction tensor $\mathbf{L}$ in $\mathbf{q} = -\mathbf{L}\mathbf{g}$ of crystalline phosphor is

$$\mathbf{L} = \begin{bmatrix} 6.44 & & \\ & 28.1 & \\ & & 83.4 \end{bmatrix} \mathrm{W/m/K}\; \mathbf{e}_i \otimes \mathbf{e}_j, \tag{13.127}$$

w. r. t. three perpendicular principle lattice directions $\mathbf{e}_i$ (Sun, 2016). We can directly determine the Voigt orientation average by applying the first isotropic projector,

$$\mathbf{L}_{\mathrm{Voigt}} = \mathbb{P}_{\mathrm{i}1} : \mathbf{L} \tag{13.128}$$

$$= \frac{\mathbf{L} : \mathbf{I}}{3} \mathbf{I} \tag{13.129}$$

$$= 39.31\bar{3}\ \mathrm{W/m/K}\, \mathbf{I}. \tag{13.130}$$

The Reuss orientation average is obtained by first inverting $\mathbf{L}$, applying $\mathbb{P}_{\mathrm{i}1}$ and inverting again,

$$\mathbf{L}_{\mathrm{Reuss}} = \left(\mathbb{P}_{\mathrm{i}1} : \mathbf{L}^{-1}\right)^{-1} \tag{13.131}$$

$$= \frac{3}{\mathbf{L}^{-1} : \mathbf{I}} \mathbf{I} \tag{13.132}$$

$$= 14.79\ \mathrm{W/m/K}\, \mathbf{I}. \tag{13.133}$$

The geometric orientation average is obtained by

$$\mathbf{L}_{\mathrm{geom}} = \exp(\mathbb{P}_{\mathrm{i}1} : \ln \mathbf{L}) \tag{13.134}$$

$$= \exp\left(\frac{\ln 6.44 + \ln 28.1 + \ln 83.4}{3}\right) \mathrm{W/m/K}\, \mathbf{I} \tag{13.135}$$

$$= 24.71\ \mathrm{W/m/K}\, \mathbf{I}. \tag{13.136}$$

In Figure 13.13, the magnitudes of heat fluxes $\mathbf{q} = -\mathbf{L} \cdot \mathbf{g}$ are depicted in dependence of the direction of the temperature gradient $\mathbf{g} = \nabla T$. The Listing 13.8 draws this plot and calculates the orientation averages.

Listing 13.8: Mathematica notebook (https://gitlab.com/gluegerainer/listings-homogenization-methods/-/blob/main/Listing-34_en.nb) for homogenizing the heat conductivity of crystalline black phosphor in case of an isotropic orientation distribution.

```
(* anisotropic conductivity of crystalline phosphor *)
L = DiagonalMatrix[{6.44, 28.1, 83.4}];
(* orientation averages according to Voigt, Reuss and geometric *)
```

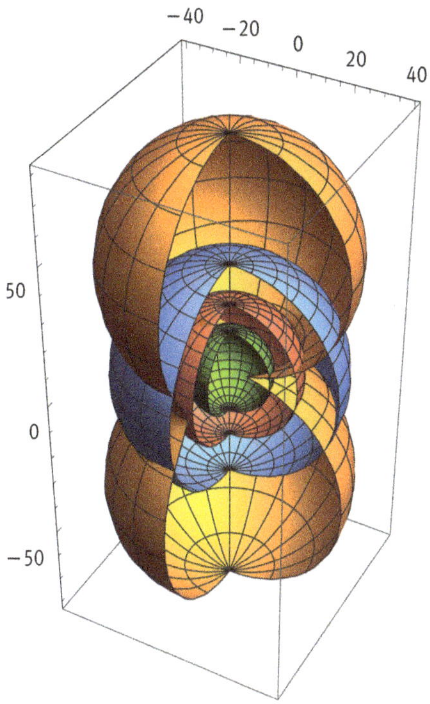

Figure 13.13: Magnitude of the heat flux $\|\mathbf{L}\cdot\mathbf{g}\|$ for anisotropic crystalline black phosphor (yellow) as well as various isotropic orientation averages (the Reuss average is green, the geometric average is red and the Voigt average is blue) over the direction of $\mathbf{g}$. The peanut shape indicates a strong anisotropy, the isotropic averages lie between the anisotropic minimum (6.44) and maximum (83.4) values.

```
LVoigt = Tr[L]/3
LReuss = 3/Tr[Inverse[L]]
LGeom = Exp[Tr[MatrixLog[L]]/3]
(* draw effective conductivities with anisotropic conductivity *)
n = {Sin[theta]*Cos[phi], Sin[theta]*Sin[phi], Cos[theta]};
SphericalPlot3D[{Norm[L.n], LVoigt, LReuss, LGeom}, {theta, 0, Pi}, {phi, 0, 3 Pi/2}]
```

13.9.2 Young's modulus of steel

Most engineers know that Young's modulus of steel is approximately 210 GPa and that Poisson's ratio is $\nu \approx 0.29$. Likewise, it is well known that metals have a grain structure and that these values are effective properties of polycrystals with more or less random orientation distributions. The single crystal stiffnesses of iron are not so well known, especially how strong their anisotropy is. Iron crystallites have a cubic crystal structure. In tensile tests, one can measure Young moduli between 132.3 GPa and 283.3 GPa, where the smallest value is obtained along the edges, and the largest value is obtained along the space diagonal. In Figure 13.14, Young's modulus is drawn as the distance from the origin over the tensile direction. One recognizes the cubic symmetry and the strong direction-dependent Young modulus (Böhlke and Brüggemann, 2001). The stiffness tetrad for an iron cubic crystal is

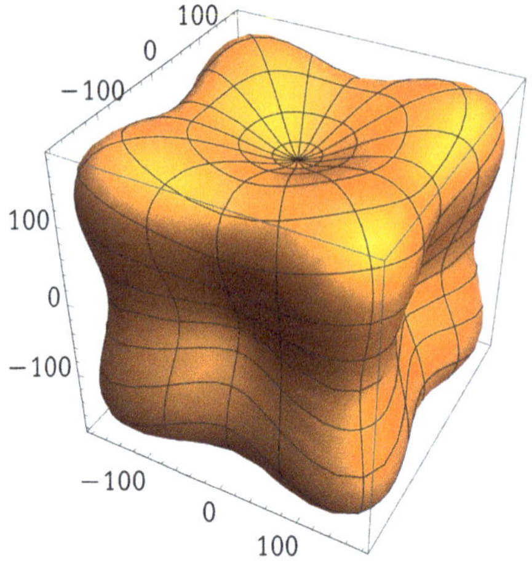

Figure 13.14: Young's modulus for a single iron crystal, depicted over the tensile direction. One sees a pronounced cubic anisotropy.

$$\mathbb{C}_{\text{Fe}} = \begin{bmatrix} 231.4 & 134.7 & 134.7 & & & \\ 134.7 & 231.4 & 134.7 & & & \\ 134.7 & 134.7 & 231.4 & & & \\ & & & 232.8 & & \\ & & & & 232.8 & \\ & & & & & 232.8 \end{bmatrix} \text{GPa } \mathbf{E}_i \otimes \mathbf{E}_j, \tag{13.137}$$

w. r. t. the normalized Voigt notation, where the lattice vectors coincide with the basis $\mathbf{e}_i$. The spectral representation is

$$\mathbb{C}_{\text{Fe}} = \underbrace{500.8}_{3K_0} \underbrace{\begin{bmatrix} \frac{1}{3} & \frac{1}{3} & \frac{1}{3} & & & \\ \frac{1}{3} & \frac{1}{3} & \frac{1}{3} & & & \\ \frac{1}{3} & \frac{1}{3} & \frac{1}{3} & & & \\ & & & & & \\ & & & & & \\ & & & & & \end{bmatrix}}_{\mathbb{P}_{C1}=\mathbb{P}_{I1}} + \underbrace{232.8}_{2G_{C2}} \underbrace{\begin{bmatrix} & & & & & \\ & & & & & \\ & & & & & \\ & & & 1 & & \\ & & & & 1 & \\ & & & & & 1 \end{bmatrix}}_{\mathbb{P}_{C2}} + \underbrace{96.7}_{2G_{C3}} \underbrace{\begin{bmatrix} \frac{2}{3} & -\frac{1}{3} & -\frac{1}{3} & & & \\ -\frac{1}{3} & \frac{2}{3} & -\frac{1}{3} & & & \\ -\frac{1}{3} & \frac{1}{3} & \frac{2}{3} & & & \\ & & & & & \\ & & & & & \\ & & & & & \end{bmatrix}}_{\mathbb{P}_{C3}}. \tag{13.138}$$

There are three cubic projectors, where the first cubic projector to the eigenvalue $3K_{\text{Fe}}$ is identical to the first isotropic projector,

$$\mathbb{P}_{I1} = \mathbb{P}_{C1}. \tag{13.139}$$

Therefore, the cubic compression modulus is invariant upon orientation averaging. Only the deviatoric part of $\mathbb{C}_{\text{Fe}}$ needs to be subjected to orientation averaging. Both cubic projectors $\mathbb{P}_{C2}$ and $\mathbb{P}_{C3}$ to the eigenvalues $2G_{C2}$ and $2G_{C3}$ give upon summation the second isotropic projector,

$$\mathbb{P}_{I2} = \mathbb{P}_{C2} + \mathbb{P}_{C3}. \tag{13.140}$$

The second cubic projector can be written symbolically as

$$\mathbb{P}_{C2} = \sum_{i=1}^{3} \mathbf{e}_i \otimes \mathbf{e}_i \otimes \mathbf{e}_i \otimes \mathbf{e}_i. \tag{13.141}$$

The isotropic projectors are discussed in Section 2.4. The scalar products among the projectors are

$$\mathbb{P}_{C1} :: \mathbb{P}_{I1} = 1 \qquad \mathbb{P}_{C2} :: \mathbb{P}_{I1} = 0 \qquad \mathbb{P}_{C3} :: \mathbb{P}_{I1} = 0 \tag{13.142}$$

$$\mathbb{P}_{C1} :: \mathbb{P}_{I2} = 0 \qquad \mathbb{P}_{C2} :: \mathbb{P}_{I2} = 3 \qquad \mathbb{P}_{C3} :: \mathbb{P}_{I2} = 2. \tag{13.143}$$

The generalized orientation average can now be written as

$$\mathbb{C}_{\text{Fe-}f} = f^{-1}\left[\int_{SO(3)} \mathbf{Q} * f(\mathbb{C}_{\text{Fe}})\, \mathrm{d}V \right] \tag{13.144}$$

$$= f^{-1}\left[\left(\mathbb{P}_{I1} \otimes \mathbb{P}_{I1} + \frac{1}{5} \mathbb{P}_{I2} \otimes \mathbb{P}_{I2} \right) :: f(\mathbb{C}_{\text{Fe}}) \right]. \tag{13.145}$$

In the following, the Voigt, Reuss and geometric orientation averages are denoted in the 6×6 matrix notation (Voigt–Mandel notation, see Section 2.3).

Voigt orientation average

The evaluation of equation (13.145) can be denoted for the Voigt average in index notation,

$$\mathbb{C}_{\text{Fe-Voigt}} = \underbrace{C_{\text{Fe}\ ij} P_{\text{I1}\ ij}}_{3K_{\text{Fe-Voigt}}} \mathbb{P}_{I1} + \underbrace{C_{\text{Fe}\ ij} P_{\text{I2}\ ij}/5}_{2G_{\text{Fe-Voigt}}} \mathbb{P}_{I2}. \tag{13.146}$$

We find

$$3K_{\text{Fe-Voigt}} = (3 \cdot 231.4/3 + 6 \cdot 134.7/3)\,\text{GPa}, \qquad K_{\text{Fe-Voigt}} = K_{\text{Fe}} = 166.93\,\text{GPa} \tag{13.147}$$

$$2G_{\text{Fe-Voigt}} = (3 \cdot 231.4 \cdot 2/3 - 6 \cdot 134.7/3 + 3 \cdot 232.8)/5\,\text{GPa} \qquad G_{\text{Fe-Voigt}} = 89.18\,\text{GPa} \tag{13.148}$$

The effective Young modulus is the inverse of the 11-, 22- or 33-component of $\mathbb{S} = \mathbb{C}_{\text{Fe-Voigt}}^{-1}$. The inversion is due to the projector representation given by the reciprocals of the eigenvalues $3K$ and $2G$,

$$E_{\text{Fe-Voigt}} = \left(\frac{1}{3} (3K_{\text{Fe-Voigt}})^{-1} + \frac{2}{3} (2G_{\text{Fe-Voigt}})^{-1} \right)^{-1}. \tag{13.149}$$

The effective Poisson ratio $\nu_{\text{Fe-Voigt}} = -\varepsilon_{\text{lateral}}/\varepsilon_{\text{axial}} = -S_{\text{Fe-Voigt }12}/S_{\text{Fe-Voigt }11}$. For the numerical values of single crystal iron given above, we obtain the effective values

$$E_{\text{Fe-Voigt}} = 227.1\,\text{GPa}, \quad \nu_{\text{Fe-Voigt}} = 0.2733. \tag{13.150}$$

Reuss orientation average

The Reuss orientation average is obtained by evaluating equation (13.145) with $f(\mathbb{C}_{\text{Fe}}) = \mathbb{C}_{\text{Fe}}^{-1}$. The inversion is easy due to the special structure of the component matrix or by using the projector representation equation (13.138). After that, we use the projectors just like we did for the Voigt mean,

$$\mathbb{S}_{\text{Fe-Reuss}} = \underbrace{[\mathbb{C}_{\text{Fe}}^{-1}]_{ij} P_{\text{I1 }ij}}_{(3K_{\text{Fe-Reuss}})^{-1}} \mathbb{P}_{\text{I1}} + \underbrace{[\mathbb{C}_{\text{Fe}}^{-1}]_{ij} P_{\text{I2 }ij}/5}_{(2G_{\text{Fe-Reuss}})^{-1}} \mathbb{P}_{\text{I2}}. \tag{13.151}$$

The reciprocal of the 11-, 22- or 33-component of $\mathbb{S}_{\text{Fe-Reuss}}$, the Poisson ratio and the shear modulus are

$$E_{\text{Fe-Reuss}} = 194.5\,\text{GPa}, \quad \nu_{\text{Fe-Reuss}} = 0.3058, \quad G_{\text{Fe-Reuss}} = 74.47\,\text{GPa}. \tag{13.152}$$

Geometric orientation average

The geometric orientation average is obtained by evaluating equation (13.145) with $f(\mathbb{C}_{\text{Fe}}) = \ln \mathbb{C}_{\text{Fe}}$. The evaluation of $\ln \mathbb{C}_{\text{Fe}}$ is again most easily done w. r. t. the eigenprojectors equation (13.138), in which the application of the logarithm is reduced to the eigenvalues. We start with

$$\ln \mathbb{C}_{\text{Fe}} = \ln(500.8\,\text{GPa})\mathbb{P}_{\text{C1}} + \ln(232.8\,\text{GPa})\mathbb{P}_{\text{C2}} + \ln(96.7)\mathbb{P}_{\text{C3}}. \tag{13.153}$$

This is projected into its isotropic part $\overset{\langle 8\rangle}{\mathbb{P}} :: \ln \mathbb{C}_{\text{Fe}}$,

$$\overset{\langle 8\rangle}{\mathbb{P}} :: \ln \mathbb{C}_{\text{Fe}} = \left(\mathbb{P}_{\text{I1}} \otimes \mathbb{P}_{\text{I1}} + \frac{1}{5}\mathbb{P}_{\text{I2}} \otimes \mathbb{P}_{\text{I2}}\right) :: \ln \mathbb{C}_{\text{Fe}} \tag{13.154}$$

$$= \ln(500.8\,\text{GPa})\mathbb{P}_{\text{I1}} + \frac{3\ln(232.8\,\text{GPa}) + 2\ln(96.7\,\text{GPa})}{5}\mathbb{P}_{\text{I2}}. \tag{13.155}$$

We finally reverse the logarithm by applying the exponential,

$$\mathbb{C}_{\text{Fe-geom}} = 500.8\text{GPa}\,\mathbb{P}_{\text{I1}} + \exp\left(\frac{3\ln(232.8) + 2\ln(96.7)}{5}\right)\text{GPa}\,\mathbb{P}_{\text{I2}} \tag{13.156}$$

$$= \underbrace{500.8\,\text{GPa}}_{3K_{\text{Fe-geom}}}\mathbb{P}_{\text{I1}} + \underbrace{163.8\,\text{GPa}}_{2G_{\text{Fe-geom}}}\mathbb{P}_{\text{I2}}, \tag{13.157}$$

from which we can extract

$$E_{\text{Fe-geom}} = 211.2\,\text{GPa}, \quad \nu_{\text{Fe-geom}} = 0.2892, \quad G_{\text{Fe-geom}} = 81.91\,\text{GPa}. \tag{13.158}$$

It is surprising how close these values are to the commonly reported effective values determined in tensile tests, which are $0.27 < \nu^* < 0.3$ and $E^* \approx 210$ GPa.

The Mathematica notebook in Listing 13.9 executes the above calculations.

Listing 13.9: Mathematica notebook (https://gitlab.com/gluegerainer/listings-homogenization-methods/-/blob/main/Listing-35_en.nb) for homogenizing the elastic properties of isotropically distributed iron single crystals. The functions `MatrixLog` and `MatrixExp` apply the logarithm and the exponential automatically in the sense we need here.

```
(* Single crystal stiffness *)
C66 = {{231.4, 134.7, 134.7, 0       ,        0, 0       },
       {134.7, 231.4, 134.7, 0       ,        0, 0       },
       {134.7, 134.7, 231.4, 0       ,        0, 0       },
       {0    , 0    , 0    , 2*116.4,        0, 0       },
       {0    , 0    , 0    , 0       , 2*116.4, 0       },
       {0    , 0    , 0    , 0       ,        0, 2*116.4}};

(* Set up projectors *)
id = {1, 1, 1, 0, 0, 0};
P166 = Outer[Times, id, id]/3;
P266 = IdentityMatrix[6] - P166;

Print["Voigt orientation average:"];
CV66 = Inner[Times, Flatten[C66], Flatten[P166]] P166 +
    Inner[Times, Flatten[C66], Flatten[P266]] P266/5;
Print["E=", EMOVOIGTISO = 1/Inverse[CV66][[1, 1]]];
Print["$\nu$=", NUVOIGTISO = -Inverse[CV66][[1, 2]]*EMOVOIGTISO];

Print["Reuss orientation average:"];
SR66 = Inner[Times, Flatten[Inverse[C66]], Flatten[P166]] P166 +
    Inner[Times, Flatten[Inverse[C66]], Flatten[P266]] P266/5;
Print["E=", EMOREUSSISO = 1/SR66[[1, 1]]];
Print["$\nu$=", NUVOIGTISO = -SR66[[1, 2]]*EMOREUSSISO];

Print["Geometric orientation average:"];
CG66 = MatrixExp[
    Inner[Times, Flatten[MatrixLog[C66]], Flatten[P166]] P166 +
     Inner[Times, Flatten[MatrixLog[C66]], Flatten[P266]] P266/5];
Print["E=", EMOGEOMISO = 1/Inverse[CG66][[1, 1]]];
Print["$\nu$=", NUVOIGTISO = -Inverse[CG66][[1, 2]]*EMOGEOMISO];
```

It gives the following output:

```
Voigt orientation average:
E=227.099277942
$\nu$=0.273263500458
Reuss orientation average:
E=194.496354874
$\nu$=0.305814342178
Geometric orientation average:
E=211.185649428
$\nu$=0.289151707839
```

13.9.3 Isotropic orientation average and the Hashin–Shtrikman bounds

We have seen in the previous section that the first cubic projector $\mathbb{P}_{C1}$ is identical to the first isotropic projector $\mathbb{P}_{I1}$, which is why all orientation averages give identical compression moduli K. Thus, orientation averaging is restricted to the deviatoric parts. This allows applying the isotropic Hashin–Shtrikman bounds to cubic polycrystals comparatively easy. We choose the comparison compression modulus $K_0 = 500.8/3\,\text{GPa}$ such that in the difference $\Delta\mathbb{C} = \mathbb{C}_{Fe} - \mathbb{C}^0$ only the shear part remains,

$$\Delta\mathbb{C} = (2G_{C2} - 2G_0)\mathbb{P}_{C2} + (2G_{C3} - 2G_0)\mathbb{P}_{C3}. \tag{13.159}$$

All inversions are restricted to the $\mathbb{P}_{I2}$ subspace, and we consider only the deviatoric parts of $\boldsymbol{\varepsilon}'$ and $\boldsymbol{\tau}'$, which is denoted by the prime. The following deviation follows the proceeding in Hashin and Shtrikman (1962a). We have the relation equation (11.1),

$$\Delta\mathbb{C}^{-1} : \boldsymbol{\tau}' = \boldsymbol{\varepsilon}'. \tag{13.160}$$

Moreover, for isotropic orientation distributions it holds

$$\tilde{\boldsymbol{\varepsilon}}' = -\mathbb{W} : \tilde{\boldsymbol{\tau}}', \tag{13.161}$$

with $\mathbb{W}$ depending only on K_0 and G_0 according to equation (12.57). We can use this to express the average deviatoric strain $\bar{\boldsymbol{\varepsilon}}' = \boldsymbol{\varepsilon}' - \tilde{\boldsymbol{\varepsilon}}'$,

$$\bar{\boldsymbol{\varepsilon}}' = \Delta\mathbb{C}^{-1} : \boldsymbol{\tau}' + \mathbb{W} : \tilde{\boldsymbol{\tau}}'. \tag{13.162}$$

We replace $\tilde{\boldsymbol{\tau}}' = \boldsymbol{\tau}' - \bar{\boldsymbol{\tau}}'$

$$\bar{\boldsymbol{\varepsilon}}' = \Delta\mathbb{C}^{-1} : \boldsymbol{\tau}' + \mathbb{W} : (\boldsymbol{\tau}' - \bar{\boldsymbol{\tau}}). \tag{13.163}$$

This can be solved for $\boldsymbol{\tau}'$,

$$\boldsymbol{\tau}' = (\Delta\mathbb{C}^{-1} + \mathbb{W})^{-1} : (\bar{\boldsymbol{\varepsilon}}' + \mathbb{W}\bar{\boldsymbol{\tau}}'). \tag{13.164}$$

We can formally take the volume average to obtain $\bar{\boldsymbol{\tau}}'$, where the homogeneous part can be factored out,

$$\bar{\boldsymbol{\tau}}' = \langle(\Delta\mathbb{C}^{-1} + \mathbb{W})^{-1}\rangle : (\bar{\boldsymbol{\varepsilon}}' + \mathbb{W} : \bar{\boldsymbol{\tau}}'). \tag{13.165}$$

The orientation average of this purely deviatoric part can be extracted with the aid of $\frac{1}{5}\mathbb{P}_{I2} \otimes \mathbb{P}_{I2}$,

$$\bar{\boldsymbol{\tau}}' = \left(\frac{1}{5}\mathbb{P}_{I2} \otimes \mathbb{P}_{I2} :: (\Delta\mathbb{C}^{-1} + \mathbb{W})^{-1}\right) : (\bar{\boldsymbol{\varepsilon}}' + \mathbb{W} : \bar{\boldsymbol{\tau}}'). \tag{13.166}$$

We insert the definitions of $\Delta\mathbb{C}$ (equation (13.159)) and

$$\mathbb{W} = w_1\mathbb{P}_{I1} + w_2\mathbb{P}_{I2} = w_1\mathbb{P}_{C1} + w_2(\mathbb{P}_{C2} + \mathbb{P}_{C3}) \tag{13.167}$$

and disregard the part that belongs to $\mathbb{P}_{C1} = \mathbb{P}_{I1}$,

$$\overline{\boldsymbol{\tau}}' = \left(\frac{1}{5}\mathbb{P}_{I2}\otimes\mathbb{P}_{I2} :: \left(\left(\frac{1}{2G_2 - 2G_0} + w_2\right)\mathbb{P}_{C2} + \left(\frac{1}{2G_3 - 2G_0} + w_2\right)\mathbb{P}_{C3}\right)^{-1}\right) : (\overline{\boldsymbol{\varepsilon}}' + \mathbb{W} : \overline{\boldsymbol{\tau}}') \tag{13.168}$$

$$= \left(\frac{1}{5}\mathbb{P}_{I2}\otimes\mathbb{P}_{I2} :: \left(\left(\frac{1}{2G_2 - 2G_0} + w_2\right)^{-1}\mathbb{P}_{C2} + \left(\frac{1}{2G_3 - 2G_0} + w_2\right)^{-1}\mathbb{P}_{C3}\right)\right) : (\overline{\boldsymbol{\varepsilon}}' + \mathbb{W} : \overline{\boldsymbol{\tau}}'). \tag{13.169}$$

The scalar products of the projectors are $\mathbb{P}_{I2} :: \mathbb{P}_{C2} = 3$ and $\mathbb{P}_{I2} :: \mathbb{P}_{C3} = 2$,

$$\overline{\boldsymbol{\tau}}' = \underbrace{\left(\frac{3}{5}\left(\frac{1}{2G_2 - 2G_0} + w_2\right)^{-1} + \frac{2}{5}\left(\frac{1}{2G_3 - 2G_0} + w_2\right)^{-1}\right)}_{B}\mathbb{P}_{I2} : (\overline{\boldsymbol{\varepsilon}}' + \mathbb{W} : \overline{\boldsymbol{\tau}}'). \tag{13.170}$$

The projector $\mathbb{P}_{I2}$ can be dropped since the right side is deviatoric, i. e., it acts as the identity. This allows identifying

$$\overline{\boldsymbol{\tau}}' = \underbrace{\frac{B}{1 - Bw_2}}_{\kappa}\overline{\boldsymbol{\varepsilon}}'. \tag{13.171}$$

This is inserted into the Hashin–Shtrikman functional equation (11.2),

$$F'(\boldsymbol{\tau}', \tilde{\boldsymbol{\varepsilon}}') = \frac{1}{2V}\int_\Omega \overline{\boldsymbol{\varepsilon}}' : \mathbb{C}^0 : \overline{\boldsymbol{\varepsilon}}' - \boldsymbol{\tau}' : \Delta\mathbb{C}^{-1} : \boldsymbol{\tau}' + \tilde{\boldsymbol{\varepsilon}}' : \boldsymbol{\tau}' + 2\boldsymbol{\tau}' : \overline{\boldsymbol{\varepsilon}}'\,\mathrm{d}V. \tag{13.172}$$

We can now replace all $\boldsymbol{\tau}$ expressions with $\overline{\boldsymbol{\varepsilon}}$ expressions, and by using equation (11.26), one can identify the Hashin–Shtrikman estimate for the effective stiffness. The bounds are obtained by inserting the extremal shear moduli.

We first insert $\boldsymbol{\varepsilon}' = \Delta\mathbb{C}^{-1} : \boldsymbol{\tau}'$,

$$F' = \frac{1}{2V}\int_\Omega \overline{\boldsymbol{\varepsilon}}' : \mathbb{C}^0 : \overline{\boldsymbol{\varepsilon}}' - \boldsymbol{\tau}' : \boldsymbol{\varepsilon}' + \tilde{\boldsymbol{\varepsilon}}' : \boldsymbol{\tau}' + 2\boldsymbol{\tau}' : \overline{\boldsymbol{\varepsilon}}'\,\mathrm{d}V, \tag{13.173}$$

followed by replacing $\boldsymbol{\varepsilon}' = \overline{\boldsymbol{\varepsilon}}' + \tilde{\boldsymbol{\varepsilon}}'$,

$$F' = \frac{1}{2V}\int_\Omega \overline{\boldsymbol{\varepsilon}}' : \mathbb{C}^0 : \overline{\boldsymbol{\varepsilon}}' - \boldsymbol{\tau}' : \overline{\boldsymbol{\varepsilon}}' - \boldsymbol{\tau}' : \tilde{\boldsymbol{\varepsilon}}' + \tilde{\boldsymbol{\varepsilon}}' : \boldsymbol{\tau}' + 2\boldsymbol{\tau}' : \overline{\boldsymbol{\varepsilon}}'\,\mathrm{d}V \tag{13.174}$$

$$F' = \frac{1}{2V}\int_\Omega \overline{\boldsymbol{\varepsilon}}' : \mathbb{C}^0 : \overline{\boldsymbol{\varepsilon}}' + \boldsymbol{\tau}' : \overline{\boldsymbol{\varepsilon}}'\,\mathrm{d}V. \tag{13.175}$$

Finally, we can replace $\boldsymbol{\tau}' = \overline{\boldsymbol{\tau}}' + \tilde{\boldsymbol{\tau}}'$. The integral over $\tilde{\boldsymbol{\tau}}' : \overline{\boldsymbol{\varepsilon}}'$ is by definition zero, such that

$$F' = \frac{1}{2V} \int_{\Omega} \overline{\boldsymbol{\varepsilon}}' : \mathbb{C}^0 : \overline{\boldsymbol{\varepsilon}}' + \overline{\boldsymbol{\tau}}' : \overline{\boldsymbol{\varepsilon}}' \, \mathrm{d}V \tag{13.176}$$

remains. Here, we can eventually drop the integration, since only homogeneous quantities appear, and replace $\overline{\boldsymbol{\tau}}' = \kappa \overline{\boldsymbol{\varepsilon}}'$ (equation (13.171)),

$$F' = \overline{\boldsymbol{\varepsilon}}' : \frac{1}{2}(\mathbb{C}_0 + \kappa)\mathbb{P}_{I2} : \overline{\boldsymbol{\varepsilon}}'. \tag{13.177}$$

With $\mathbb{C}_0 = 3K_0\mathbb{P}_{I1} + 2G_0\mathbb{P}_{I2}$ and the restriction to the deviatoric part only,

$$F' = \underbrace{[G_0 + \kappa/2]}_{G_{HS}} \overline{\boldsymbol{\varepsilon}}' : \overline{\boldsymbol{\varepsilon}}' \tag{13.178}$$

remains, with the Hashin–Shtrikman-shear modulus G_{HS}. Summarizing this expression for G_{HS} gives

$$G_{HS} = \frac{3K_0(9G_0G_{C2} + 6G_0G_{C3} + 10G_{C2}G_{C3}) + 4G_0(6G_0G_{C2} + 4G_0G_{C3} + 15G_{C2}G_{C3})}{3K_0(15G_0 + 4G_{C2} + 6G_{C3}) + 4G_0(10G_0 + 6G_{C2} + 9G_{C3})}. \tag{13.179}$$

We obtain the Hashin–Shtrikman bounds by replacing G_0 either by G_{C2} and G_{C3},

$$G_{HS-} = \frac{G_{C3}(G_{C2}(84G_{C3} + 57K_0) + 2G_{C3}(8G_{C3} + 9K_0))}{12G_{C2}(2G_{C3} + K_0) + G_{C3}(76G_{C3} + 63K_0)} \tag{13.180}$$

$$G_{HS+} = \frac{G_{C2}(24G_{C2}^2 + 76G_{C2}G_{C3} + 27G_{C2}K_0 + 48G_{C3}K_0)}{64G_{C2}^2 + 36G_{C2}G_{C3} + 57G_{C2}K_0 + 18G_{C3}K_0}. \tag{13.181}$$

The integer coefficients result from the products of the projectors and the coefficients in w_2. For iron, one finds the bounds

$$80.851\,\text{GPa} < G^* < 83.448\,\text{GPa}, \tag{13.182}$$

which can be converted into Young's modulus by $E_{HS\pm} = 9K_0G_{HS\pm}/(3K_0 + G_{HS\pm})$,

$$208.84\,\text{GPa} < E^* < 214.59\,\text{GPa}. \tag{13.183}$$

One can see that the bounds are much stricter than the Voigt–Reuss bounds. The Mathematica notebook in Listing 13.10 carries out the above calculations.

Listing 13.10: Mathematica notebook (https://gitlab.com/gluegerainer/listings-homogenization-methods/-/blob/main/Listing-37_en.nb) for the Hashin–Shtrikman bounds for the isotropic orientation distribution of cubic single crystals, evaluated for iron.

```
Remove["Global`*"]
w2 = 3 (K0 + 2 G0)/(5 G0 (3 K0 + 4 G0));
B = 3/5 /(1/(2 GC2 - 2 G0) + w2) + 2/5/(1/(2 GC3 - 2 G0) + w2);
kappa = B/(1 - B w2);
GHS = G0 + kappa/2 // FullSimplify
GHSminus = FullSimplify[ GHS /. {G0 -> GC3}] (* For iron GC3 < GC2, therefore "minus" *)
GHSplus =  FullSimplify[GHS /. {G0 -> GC2}]
(* Single crystal constants for iron *)
K0 = 500.8/3;  (* usually (c11+2 c12)/3 *)
GC2 = 116.4;   (* usually 2 c44 /2 when a nonnormalized basis is used *)
GC3 = 48.36;   (* usually  (c11-c12)/2 *)
Print["Bounds for the shear modulus: ", GHSminus , " < G < ", GHSplus]
EHSminus = 9 K0 GHSminus/(3 K0 + GHSminus);
EHSplus = 9 K0 GHSplus/(3 K0 + GHSplus);
Print["Bounds for Young's modulus: ", EHSminus , " < E < ", EHSplus]
```

Bibliography

Aleksandrov, K. S. and L. A. Aisenberg (1966). "Method of calculating physical constants of polycrystalline metals". *Soviet Physics. Doklady* 11.3, pp. 323–325. 56

Andrianov, Igor V., Galina A. Starushenko, and Vladimir A. Gabrinets (2018). "Percolation threshold for elastic problems: Self-consistent approach and Padé approximants". In: *Advances in Mechanics of Microstructured Media and Structures*. Ed. by Francesco dell'Isola, Victor A. Eremeyev, and Alexey Porubov. Springer International Publishing, pp. 35–42. 84

Auffray, N., Q. C. He, and H. Le Quang (2019). "Complete symmetry classification and compact matrix representations for 3D strain gradient elasticity". *International Journal of Solids and Structures* 159, 197210. 151

Bertram, A. and R. Glüge (2015). *Solid Mechanics*. Springer. ISBN: 978-3-319-19566-7. 7, 43

Böhlke, T. (2005). "Application of the maximum entropy method in texture analysis". *Computational Materials Science* 32, pp. 276–283. 165

Böhlke, T. and C. Brüggemann (2001). "Graphical representation of the generalized Hooke's law". *Technische Mechanik* 21, pp. 145–158. 173

Boussinesq, J. (1885). *Application des potentiels à l'étude de l'équilibre et du mouvement des solides élastiques*. Gauthier-Villars, Paris. 59

Brannon, R. M. (2018). *Rotation, Reflection, and Frame Changes*. IOP Publishing. 2, 7, 141, 148, 166

Bunge, H. J. (1965). "Eine Bemerkung zur Darstellung von Blechtexturen durch drei inverse Polfiguren". *Zeitschrift für Metallkunde* 56.6, pp. 378–379. 165

Bunge, H. J. (1969). *Mathematische Methoden der Texturanalyse*. Akademie-Verlag. 162

Chen, Y. (2008). *Percolation and Homogenization Theories for Heterogeneous Materials*. Ph. D. thesis at Massachusetts Institute of Technology. 81, 82, 84

Cowin, S. C. and M. M. Mehrabadi (1992). "The structure of the linear anisotropic elastic symmetries". *Journal of the Mechanics and Physics of Solids* 40.7, pp. 1459–1471. 7, 153

Curie, P. (1894). "Sur la symétrie dans les phénomènes physiques, symétrie d'un champ électrique et d'un champ magnétique". *Journal de Physique* 3, pp. 393–415. 151

deBotton, G. (2005). "Transversely isotropic sequentially laminated composites in finite elasticity". *Journal of the Mechanics and Physics of Solids* 53.6, pp. 1334–1361. 59

Efendiev, Y. and T. Y. Hou (2009). *Multiscale Finite Element Methods: Theory and Applications*. Surveys and Tutorials in the Applied Mathematical Sciences. Springer New York. 38

Eshelby, J. D. (1957). "The determination of the elastic field of an ellipsoidal inclusion, and related problems". *Proceedings of the Royal Society of London. Series A: Mathematical, Physical and Engineering Sciences* 241, pp. 376–396. 59, 71

Feyel, F. (1999). "Multiscale FE^2 elastoviscoplastic analysis of composite structures". *Computational Materials Science* 16.1–4, pp. 344–354. 37

Francfort, G. and F. Murat (1986). "Homogenization and optimal bounds in linear elasticity". *Archive for Rational Mechanics and Analysis* 94.4, pp. 307–334. 66

Fritzen, F. and T. Böhlke (2010a). "Influence of the type of boundary conditions on the numerical properties of unit cell problems". *Technische Mechanik* 30.4, pp. 354–363. 38

Fritzen, F. and T. Böhlke (2010b). "Three-dimensional finite element implementation of the nonuniform transformation field analysis". *International Journal for Numerical Methods in Engineering* 84, pp. 803–829. 37

Garboczi, E. J. et al. (1995). "Geometrical percolation threshold of overlapping ellipsoids". *Physical Review E* 52.1, pp. 819–828. 82, 83

Gibson, L. J. and M. F. Ashby (1999). *Cellular Solids: Structure and Properties*. Cambridge Solid State Science Series. Cambridge University Press. 39

https://doi.org/10.1515/9783110793529-014

Glüge, R. (2013). "Generalized boundary conditions on representative volume elements and their use in determining the effective material properties". *Computational Materials Science* 79, pp. 408–416. 36, 38

Glüge, R. (2016). "Effective plastic properties of laminates made of isotropic elastic plastic materials". *Composite Structures* 149, pp. 434–443. 59

Glüge, R. (2018). "Principles of material modeling". In: *Encyclopedia of Continuum Mechanics*. Ed. by Holm Altenbach and Andreas Öchsner. Berlin, Heidelberg: Springer Berlin Heidelberg, pp. 1–8. ISBN: 978-3-662-53605-6. 48

Glüge, R. and M. Weber (2013). "Numerical properties of spherical and cubical representative volume elements with different boundary conditions". *Technische Mechanik* 33.2, pp. 97–103. 38

Glüge, R., M. Weber, and A. Bertram (2012). "Comparison of spherical and cubical statistical volume elements with respect to convergence, anisotropy, and localization behavior". *Computational Material Science* 63, pp. 91–104. DOI: 10.1016/j.commatsci.2012.05.063. 37, 38

Glüge, R. et al. (2020). "On the difference between the tensile stiffness of bulk and slice samples of microstructured materials". *Applied Composite Materials* 27, pp. 969–988. 28

Gross, D. and T. Seelig (2011). *Fracture Mechanics with an Introduction to Micromechanics*, 2. Edition. Springer. ISBN: 978-3-642-19240-1. 2, 59, 78, 81

Halphen, Bernard and Quoc Son Nguyen (1975). "Sur les matériaux standard généralisés". *Journal de Mécanique* 14, pp. 39–63. 33

Hashin, Z. (1983). "Analysis of composite materials – a survey". *Journal of Applied Mechanics* 50, 481505. 26

Hashin, Z. and S. Shtrikman (1962a). "A variational approach to the theory of the elastic behaviour of polycrystals". *Journal of the Mechanics and Physics of Solids* 10.4, pp. 343–352. ISSN: 0022-5096. 178

Hashin, Z. and S. Shtrikman (1962b). "On some variational principles in anisotropic and nonhomogeneous elasticity". *Journal of the Mechanics and Physics of Solids* 10.4, pp. 335–342. 85

Hashin, Z. and S. Shtrikman (1963). "A variational approach to the theory of the elastic behaviour of multiphase materials". *Journal of the Mechanics and Physics of Solids* 11.2, pp. 127–140. 85, 91, 95

Hazanov, S. and C. Huet (1994). "Order relationships for boundary condition effects in heterogeneous bodies smaller than the representative volume". *Journal of the Mechanics and Physics of Solids* 42, pp. 1995–2011. 38

He, Q.-C. and Z.-Q. Feng (2012). "Homogenization of layered elastoplastic composites: theoretical results". *International Journal of Non-Linear Mechanics* 47.2, pp. 367–376, Nonlinear Continuum Theories. 60

Helnwein, P. (2001). "Some remarks on the compressed matrix representation of symmetric second-order and fourth-order tensors". *Computer Methods in Applied Mechanics and Engineering* 190.22, pp. 2753–2770. 7

Hermann, C. (1934). "Tensoren und Kristallsymmetrie". *Zeitschrift für Kristallographie – Crystalline Materials* 89, pp. 32–48. 152

Hill, R. (1963). "Elastic properties of reinforced solids: Some theoretical principles". *Journal of the Mechanics and Physics of Solids* 11.5, pp. 357–372. 31

Hill, R. (1964). "Theory of mechanical properties of fibre-strengthened materials: I. Elastic behaviour". *Journal of the Mechanics and Physics of Solids* 12.4, pp. 199–212. 59

Hill, R. (1983). "Interfacial operators in the mechanics of composite media". *Journal of the Mechanics and Physics of Solids* 31.4, pp. 347–357. 60

Houdaigui, F. et al. (2007). "On the size of the representative volume element for isotropic elastic polycrystalline copper". In: *IUTAM Symposium on Mechanical Behavior and Micro-Mechanics of Nanostructured Materials*. Ed. by Y. L. Bai, Q. S. Zheng, and Y. G. Wei. Solid Mechanics and its Applications, Vol. 144. Springer Netherlands, pp. 171–180. 37

Huet, C. (1990). "Application of variational concepts to size effects in elastic heterogeneous bodies". *Journal of the Mechanics and Physics of Solids* 38, pp. 813–841. 38

Kalisch, J. and R. Glüge (2015). "Analytical homogenization of linear elasticity based on the interface orientation distribution – a complement to the self-consistent approach". *Composite Structures* 126, pp. 398–416. 98

Kirsch, E. G. (1898). "Die Theorie der Elastizität und die Bedürfnisse der Festigkeitslehre". *Zeitschrift Des Vereines Deutscher Ingenieure* 42, pp. 797–807. 59

Klusemann, B. and M. Ortiz (2015). "Acceleration of material-dominated calculations via phase-space simplicial subdivision and interpolation". *International Journal for Numerical Methods in Engineering* 103.4, pp. 256–274. 37

Kowalczyk-Gajewska, K. and J. Ostrowska-Maciejewska (2009). "Review on spectral decomposition of Hookes tensor for all symmetry groups of linear elastic material". *Engineering Transactions* 57.3–4, pp. 145–183. 152

Laws, N. (1975). "On interfacial discontinuities in elastic composites". *Journal of Elasticity* 5.3–4, 227235. 60

Lee, C. E., A. E. Ozdaglar, and D. Shah (2014). "Solving systems of linear equations: Locally and asynchronously". *CoRR*, abs/1411.2647. 125

Liu, L. and Z. Huang (2014). "A note on Mori-Tanaka's method". *Acta Mechanica Solida Sinica* 27.3, pp. 234–244. 74

Llorca, J., C. González, and J. Segurado (2007). "5 – Finite element and homogenization modelling of materials". In: *Multiscale Materials Modelling*. Ed. by Z. Xiao Guo. Woodhead Publishing Series in Civil and Structural Engineering. Woodhead Publishing, pp. 121–147. 38

Mandel, J. (1966). "Contribution théorique à l'étude de l'écrouissage et des lois de l'écoulement plastique". In: *Applied Mechanics*. Ed. by Henry Görtler. Berlin, Heidelberg: Springer Berlin Heidelberg, pp. 502–509. ISBN: 978-3-662-29364-5. 31

Matthies, S. and M. Humbert (1995). "On the principle of a geometric mean of even-rank symmetric tensors for textured polycrystals". *Journal of Applied Crystallography* 28.3, pp. 254–266. 56

McCurdy, A. K., H. J. Maris, and C. Elbaum (1970). "Anisotropic heat conduction in cubic crystals in the boundary scattering regime". *Physical Review B* 2.10, pp. 4077–4083. 152

Michel, J. C. and P. M. Suquet (2003). "Nonuniform transformation field analysis". *International Journal of Solids and Structures* 40, pp. 6937–6955. 37

Milton, G. W. (2002). *The Theory of Composites*. Cambridge University Press. 2, 19, 21, 59, 85, 115, 139

Morawiec, A. (2004). *Orientations and Rotations – Computations in Crystallographic Textures*. Springer. 2, 141, 158

Moulinec, H. and P. Suquet (1998). "A numerical method for computing the overall response of nonlinear composites with complex microstructure". *Computer Methods in Applied Mechanics and Engineering* 157.1, pp. 69–94. 131

Mura, T. (1987). *Micromechanics of Defects in Solids*. Mechanics of Elastic and Inelastic Solids, Vol. 3. Springer Netherlands. 71, 139

Neumann, F. E. (1885). *Vorlesungen über die Theorie der Elastizität der festen Körper und des Lichtäthers*. Ed. by O. E. Meyer. Teubner-Verlag Leipzig. 151

Nomura, S. (2016). *Micromechanics with Mathematica*. Wiley. 2

Nygards, M. (2003). "Number of grains necessary to homogenize elastic materials with cubic symmetry". *Mechanics of Materials* 35, pp. 1049–1053. 37

Orera, Victor M. and Rosa I. Merino (2015). "Ceramics with photonic and optical applications". *Boletín de la Sociedad Española de Cerámica y Vidrio* 54.1, pp. 1–10. 18
Popko, E. S. (2012). *Divided Spheres: Geodesics and the Orderly Subdivision of the Sphere*. Taylor & Francis. 141
Reuss, A. (1929). "Berechnung der Fließgrenze von Mischkristallen auf Grund der Plastizitätsbedingung für Einkristalle". *ZAMM - Journal of Applied Mathematics and Mechanics / Zeitschrift fur Angewandte Mathematik und Mechanik* 9.1, pp. 49–58. 51
Roe, Ryong-Joon (1965). "Description of crystallite orientation in polycrystalline materials. III. General solution to pole figure inversion". *Journal of Applied Physics* 36.6, pp. 2024–2031. 165
Schröder, J., D. Balzani, and D. Brands (2011). "Approximation of random microstructures by periodic statistically similar representative volume elements based on lineal-path functions". *Archive of Applied Mechanics* 81 (7), pp. 975–997. 38
Schröder, Jörg (2014). "A numerical two-scale homogenization scheme: the FE^2-method". In: *Plasticity and Beyond: Microstructures, Crystal-Plasticity and Phase Transitions*. Ed. by Jörg Schröder and Klaus Hackl. Vienna: Springer Vienna, pp. 1–64. 37
Sun, B. et al. (2016). "Temperature dependence of anisotropic thermal-conductivity tensor of bulk black phosphorus". *Advanced Materials* 29.3, 1603297. 172
Zohdi, T. I. and P. Wriggers (2008). *An Introduction to Computational Micromechanics*. Lecture Notes in Applied and Computational Mechanics. Springer, Corrected Second Printing. 38
Tashkinov, M. (2017). "Statistical methods for mechanical characterization of randomly reinforced media". *Mechanics of Advanced Materials and Modern Processes* 3.1, pp. 1–18. 20
Thomson, W. (1856). "XXI. Elements of a mathematical theory of elasticity". *Philosophical Transactions of the Royal Society of London* 146, pp. 481–498. 7
Ting, T. C. T. (1996). *Anisotropic Elasticity: Theory and Applications*. Oxford University Press. 153
Torquato, S. (1997). "Effective stiffness tensor of composite media – I. Exact series expansions". *Journal of the Mechanics and Physics of Solids* 45.9, pp. 1421–1448. 66, 139
Torquato, S. (2005). *Random Heterogeneous Materials: Microstructure and Macroscopic Properties*. Interdisciplinary Applied Mathematics. Springer New York. ISBN: 978-0-387951-67-6. 19, 21
Voigt, W. (1889). "Über die Beziehung zwischen den beiden Elastizitatskonstanten isotroper Körper". *Annalen der Physik* 274.12, pp. 573–587. 51
Voigt, W. (1928). *Lehrbuch Der Kristallphysik*. B. G. Teubners Sammlung von Lehrbüchern auf d. Geb. d. math. Wiss. Johnson Reprint Corporation. ISBN: 978-0-384648-40-1. 51
Watt, J. P. and L. Peselnick (1980). "Clarification of the Hashin-Shtrikman bounds on the effective elastic moduli of polycrystals with hexagonal, trigonal, and tetragonal symmetries". *Journal of Applied Physics* 51.3, pp. 1525–1531. 85
Weyl, H. (1939). *The Classical Groups: Their Invariants and Representations*. Princeton Mathematical Series. Princeton University Press. ISBN: 0-691-07923-4. 151
Wigner, E. P. (1931). *Gruppentheorie und ihre Anwendung auf die Quantenmechanik der Atomspektren*. J. W. Edwards. 162
Wulfinghoff, S. and S. Reese (2016). "Efficient computational homogenization of simple elastoplastic microstructures using a shear band approach". *Computer Methods in Applied Mechanics and Engineering* 298, pp. 350–372. 37
Yanase, Keiji (Jan. 2019). *A Derivation of Eshelby's Tensor for a Spherical Inclusion*. 71
Yvonnet, J. (2019). *Computational Homogenization of Heterogeneous Materials with Finite Elements*. Solid Mechanics and Its Applications. Springer International Publishing. 38

Index

https://doi.org/10.1515/9783110793529-015

www.ingramcontent.com/pod-product-compliance
Lightning Source LLC
Chambersburg PA
CBHW081524220326
41598CB00036B/6326